Integrated Circuits
Theory and Applications

Integrated Circuits
Theory and Applications

Charles F. Wojslaw

San Jose City College

RESTON PUBLISHING COMPANY, INC.
A Prentice-Hall Company
Reston, Virginia

PHOTO CREDITS:

Figures 1-3, 2-3, 2-9, 2-10, and 2-12 courtesy of National
  Semiconductor Corporation, Santa Clara, California.

Figures 2-2 and 2-8 courtesy of RCA Solid State Division,
  Somerville, New Jersey.

Library of Congress Cataloging in Publication Data

Wojslaw, Charles F.
  Integrated circuits.

  Includes index.
  1. Integrated circuits. I. Title.
TK7874.W63   621.381'73   77-27300
ISBN 0-87909-379-X

© 1978 by Reston Publishing Company, Inc.
A Prentice-Hall Company
Reston, Virginia 22090

10 9 8 7 6 5 4 3 2 1

Printed in the United States of America

to my God
to my country
to my family

Contents

Preface

The electronics industry is becoming an important segment of our national economy. The movement of electronic products into the consumer, business, and industrial markets has dramatically increased the growth of the industry and sales of its equipment. This growth is primarily attributed to the integrated circuit, or IC.

The underlying technology of the electronics industry is "solid state." Knowledge of the technology and its prime product, integrated circuits, is now required not only by IC manufacturing people but is essential for all electronics personnel. From assembler to administrator, from circuit to system, from design to field service, the IC touches all phases of electronics. Most major original equipment manufacturers (OEM) now have internal IC fabrication capabilities, and the economic success of these companies greatly depends upon their ability to translate the equipment's circuits to integrated form. The success of these companies also depends upon the solid state expertise of their personnel.

The breadth and depth of integrated circuit technology prohibits a comprehensive coverage of the subject in one book. It would be impossible to cover the electronic aspects of all the devices, their applications, and the associated physics and chemistry of all the manufacturing processes.

The intent of this book is to cover integrated circuits from their fabrication to their application. To do justice to the topics, only those processes and devices that presently have the greatest impact in the industry are discussed. This impact is measured primarily by the number of devices that are sold in each area. The application of these devices is limited to the fundamental circuits, which sufficiently illustrates the value of the devices as components used to achieve more complex functions.

Integrated circuits may be classified according to how they are made (bipolar or unipolar) or how they are used (linear or digital). This book explores both areas and covers the dominant members of each class. In this way, the reader may achieve a broad, fundamental knowledge of integrated circuits that may later be developed in specialized areas.

Before one can understand the whole, he must know the parts. Before one can understand the parts, he must know how they are made. And before all of this, he must be convinced that this understanding he seeks is worth his while and is important.

Chapter 1 provides an overview of the electronic products that utilize integrated circuits and where they are used. In addition, key terms are defined, the IC classifications are diagrammed, and the IC features are discussed.

Chapter 2 takes the fabrication of the IC from the raw material, through all the intermediate steps, to the final product. The bipolar manufacturing process is emphasized.

Chapter 3 identifies the key integrated components and their structural form. Bipolar and unipolar (MOS) components are described.

Chapters 4 and 5 cover the two levels of complexity of bipolar digital devices. TTL devices are highlighted, but other related device types are introduced.

Chapter 6 spans unipolar or MOS digital devices. CMOS is highlighted in the simpler devices, while p and n MOS are used to illustrate the complex system-level devices.

Chapters 7 and 8 are devoted to the four dominant types of linear devices. Chapter 7 concentrates on the operational and current-mode amplifiers, while Chapter 8 deals with positive and negative voltage regulators, and comparators.

I would like to thank all my colleagues from the industry and from college because their association has always been a learning experience. But most of all, I would like to thank my wife for providing an atmosphere at home of one of love and happiness.

CHARLES F. WOJSLAW

Integrated Circuits
Theory and Applications

Integrated Circuits

The integrated circuit, or IC, is a miniature, low-cost electronic component that performs a high-level function. This chapter discusses the role of the IC in man's life, explores its characteristics, and defines key terms associated with the IC and its technology.

1.1. THE ROLE OF THE IC

If one were to walk up to a person on the street and say, "What role does electronics play in your life?" he would probably pause for a moment and state that except for his radio and television, its role is small. He couldn't be further from the truth. Most people, including electronics personnel, do not fully appreciate the tremendous impact that electronics has in our day-to-day living.

Integrated circuits are electronics, and today electronics is primarily integrated circuits. The story of one cannot be told without the other. This statement can be challenged, successfully, by a few segments of the industry, but overall the present growth and effect of electronics is traced back to the miniature, low-power, reliable, cost-effective, well-performing integrated circuit.

An interesting footnote to the story of the integrated circuit is the fact that it was developed in 1958. This rather noteworthy technical achievement has occurred quite recently.

Many people contribute to the ultimate success of a technical project. The development of the IC was no exception. However, two men were prominent in the development of the IC as a viable electronic component. Jack Kilby, at Texas Instruments, was credited with conceiving and constructing the first working monolithic circuit in 1958. Robert Noyce was cited for his sophistication of the monolithic circuit for more specialized use, particularly in the industry. From the efforts of these men and their colleagues, a technology has emerged which today supports a dozen major manufacturers whose products touch everyone.

The most obvious area that highlights electronic IC products is the consumer area. When one talks about the *volume selling* of a product, he is talking about the consumer area. For years people have balanced their checkbooks and performed simple arithmetic operations using pencil and paper; doing so isn't so time-consuming or difficult. Yet, today, millions and millions of people own and use the electronic calculator. It saves a few minutes of time and is highly economical and convenient to use. All mathematical operations are performed by a single IC device.

Mechanical timepieces, such as clocks and watches, are proven products. They are economical and reliable. Yet the precision of numbers, the additional reliability and slightly lower cost, and the attractiveness of luminescent displays are prompting people to buy digital clocks and watches. The watch industry is indeed moving from abroad to the silicon valleys of the IC manufacturers.

Nearly everyone owns a radio and a television set, and they too are improving. Television advertises solid-state TV's; in a few cases, they mean transistors, but new models use solid-state ICs. Lower power consumption, electronic tuning, built-in clocks, larger viewing screens, and game attachments are a few of the recent features added to enhance the value and pleasure of this communications receiver. Are radio and TV the extent of the entertainment electronics we have? Definitely not. High-fidelity stereos, TV games. FM receivers, electronic organs, cameras, jukeboxes, magnetic tape recorders, pinball machines, toys, and video disc players have to be added to the list.

Still close to home we have to look at the automobile and appliances. The need for greater economy, reliability, efficiency, and control has led automobile manufacturers to look to electronics for solutions. Digital clocks, seat belt alarms, electronic tachometers and ignitions, and controls for pollution, fuel, speed, and the drive train are being added to most new cars. Electronic timers and controls for clothes dryers, ranges, sewing machines, washing machines, and other appliances increase the quality of these home time-savers.

In the field of transportation, the automobile is not alone in reaping the benefits of the IC. Computer-controlled mass transit systems with their electrically operated cars offer an alternative to the automobile in urban areas. Large, modern airplanes replaced their hydraulic controls with electrically operated controls and now use computers to monitor the myriad of airborne subsystems. Communication and navigational systems are still electronic but are now highly advanced. The small size and low weight of ICs have provided the extra capability required without a significant increase in weight and space.

The effect of computers on our day-to-day living is profound. Computers come in three sizes: small, medium, and large; but it has been the small computer that has created the uproar. The power of the one-chip computer called a microprocessor has prompted all businesses and industries to adapt it to their applications. Production is increased and routine, boring human tasks are eliminated through the microprocessor-based process control equipment found in mills, refineries, and printers.

A trip to the local, chain-operated grocery store will reveal the effort of businesses in seeking more cost-effective solutions using computer power. More than likely, the clerk at the store will tabulate your bill using an electronic, point-of-scale (POS) cash register. Unnoticed, your purchases are transmitted to a central computer in the back room, where a continuous, up-to-date log is maintained on the store's inventory and sales. Inventory is continuously ongoing and highly accurate. Supplies are requisitioned at the moment that the available quantity falls below a predetermined amount. Furthermore, management is fully cognizant of the breadth, depth, and time of sales of its business. Many retail stores are sure to follow.

All moderately large companies use computers to generate the payroll. The neatly typed paychecks we receive may appear impersonal, but they represent a step in progress from which we benefit. Generating a payroll is but one of many tasks performed by the large data processing computers found in business and industry. It would be economically impractical to maintain the money transactions at banks, the sale of stocks at the exchanges, and the goods inventory at large stores without the computer. Economy is not the only benefit reaped. Accuracy, speed, and efficiency make the computer an invaluable business and industrial tool.

Then does the computer eliminate jobs? No. More jobs are created, but the job skill level shifts. The need for manual laborers performing repetitive tasks is being replaced by the need for skilled operators, for people of interdisciplinary backgrounds, and for those who perform service functions.

The heartbeat of the world is continuously monitored by the communications media. No longer are countries and their peoples just dots on a map. They are as close as our living rooms. Low-power, lightweight electronic communications satellites relay the events of the world to us as they happen. Visual communications are not limited to television. Sections of the country are already using video telephones, and only the cost of replacing older systems is hindering their widespread usage. The popularity of citizen's band (CB) transceivers began with the ham operators. Truck drivers adapted it for truck-to-truck communication, and now, large scale car-to-car and car-to-home CB transmission is imminent. Children mimic their parents using walkie-talkies. Answering service monitors, Dictaphones, typewriters, and reproduction machines add to the list. The electronic black boxes associated with lasers and fiber-optics further push progress in the communications field.

People, as they live better, want to live longer. The medical profession removes cataracts, seals wounds, and burns cancerous tissue using lasers. Micro-miniature pacemakers help people's hearts beat. Electrocardiograms, electroencephalograms, and other diagnostic tests are used by doctors to confirm the functioning of vital body functions. Critically ill patients are monitored in intensive care units by sophisticated electronic equipment. Fluid analyzers, fluid processing machines, X-ray scanners, and prosthetic devices are added to the number of products where electronics and integrated circuits have helped to aid the medical profession.

The center of the educational system is still the student and the teacher. Learning begins with teaching and continues forever through studying. However, this procedure today is supplemented with closed-circuit TV (CCTV), video tape players, interactive man-machine interfaces, film strips, and other audio-visual aids.

The hush-hush military and aerospace industries are at the forefront in using the sophistication and power of electronics. The communications, navigation, radar, sonar, and fire control systems of military equipment and vehicles utilize every known advantage that ICs offer. So does our successful aerospace program.

Will ICs and the industry of which they are a part continue to extend their influence and enhance mankind's livelihood? Until something comes along that is better, yes. Probably more important than its present accomplishments are the industry's future and potential. However, one person's imagination is insufficient to foresee the myriad of products that the industry will produce to push the progress of man just one step further.

1.2. DEFINITIONS

Every science, and even its branches, has a vocabulary that is unique in describing its parts and principles. This section defines those key words that are necessary for the understanding of integrated circuits.

Integrated Circuit. A *circuit* is a collection of components connected together in a unique and complete configuration to perform some useful electronic function. An *integrated circuit*, or IC, is a circuit whose components are manufactured on a continuous or single piece of semiconductor material. If all of the components are on a single piece of material (or chip), the integrated circuit is called *monolithic*, which literally means *one-stone*. If two or more interconnected chips or components comprise the circuit, the IC is said to be *hybrid*.

Solid State. The control of current by electronic devices that are continuous in structure and thus contain no moving parts, filaments, or vacuum gaps. Examples include crystals, transistors, and integrated circuits.

Solid State Technology. The science of producing solid state devices.

Planar Process. The basic manufacturing process of solid state technology. It derives its name from the fact that all integrated circuit components are interconnected on the same surface or *plane* of the semiconductor material. This process embodies four key principles: (1) diffusion, (2) oxidation, (3) selective oxide removal, and (4) epitaxy.

Diffusion. The intermingling of the atomic structures of two materials. The penetration or permeation of a gas or liquid in a solid material. Examples include water (or steam) and wood, ink and paper, and a phosphorous gas and silicon.

Oxidation. The conversion of an element to a new element through its combination with oxygen. Examples include ferric oxide (rust) and silicon dioxide.

Selective Oxide Removal. The ability to remove an element containing oxygen, i.e. silicon dioxide, from a selected area.

Deposition. The act of precipitating or laying down by a natural process. The procedure of evaporating a metal, usually aluminum, and precipitating or depositing it on a surface, usually silicon, to form metallic interconnections and pads for IC components.

Wafer. A slice or circular disc of purified silicon. It is usually 20 mils thick and three to four inches in diameter.

Die. A portion of a wafer containing an integrated circuit.

Dopant. An element, or impurity, added to silicon to modify the silicon's electrical properties. Examples include boron and phosphorous.

Epitaxy. The growing of one crystalline material on another crystalline material. The two materials have similar physical structures but are chemically different.

1.3. CLASSIFICATIONS

The two general fabrication categories of ICs, shown in Figure 1-1, include the monolithic and hybrid types. The number of specific manufacturing processes within these two categories is large, and each one is individually unique and complex. The greatest number of devices sold today are *bipolar* and *unipolar* monolithic ICs. To this category we exclusively direct our attention.

Bipolar ICs are those devices whose components are *current controlled* and require both positive (+) and negative (−) polarity currents for operation in

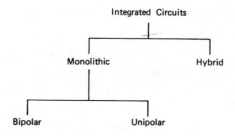

Figure 1-1. Classification of Integrated Circuits by Fabrication

their active elements. Active elements refer to those leads of a device that directly affect the input/output characteristics for that device. An example of a bipolar component is the npn bipolar junction transistor. This is the original, common transistor.

Unipolar monolithic ICs pertain to those devices whose components are primarily *voltage controlled* and have a single polarity operational current in its active elements. An example of a unipolar component is the MOS (*Metal-Oxide-Semiconductor*) transistor.

The two general functional categories of ICs shown in Figure 1-2 are of the digital and linear types. Although the present trend in IC technology is to build increasinly more complex circuits, some of which do both, the great majority of devices perform either digital *or* linear functions. Neither category is the more important, since each type does a unique job.

Digital ICs are those devices whose inputs and outputs have two discrete states; *linear* ICs are those devices whose inputs and outputs are proportionally or mathematically related.

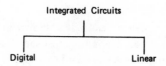

Figure 1-2. Classification of Integrated Circuits by Function

1.4. FEATURES

Integration refers to the manufacturing process that combines separate electronic components into an integral whole or single chip. This process possesses the advantage of being able to produce numerous devices at one time and is called *batch processing*. The control and uniformity within the process allow devices to be produced at a low unit cost.

In addition to cost, the IC has small size and weight, is reliable, and consumes a small amount of power. These features have led to countless applications in the military, aerospace, consumer, business, and industrial fields. The success of recent aerospace endeavors would not be possible without the high reliability and small size of electronic components in communications and control equipment. Battery-operated consumer products, such as calculators and watches, depend on low power consumption to maximize the operating life of the product. The reduction in size and the increased reliability of digital computers have led to their application in nonelectronic industries and business. Of course, low cost has pushed the usage of electronics in all fields.

The IC is a development of solid state technology. However, this technology has also spawned transducers, sensors, and optoelectronic devices. Figure 1-3 illustrates the many sizes and shapes of semiconductor devices and products.

Figure 1-3. Semiconductor Devices and Products

Integrated Circuit Fabrication

The integrated circuit is the result of a multistep physical and chemical process. This chapter discusses the procedural steps in converting raw material to a complex electronic device.

2.1. INTRODUCTION

The manufacture of integrated circuits is an extremely complex process. This process is radically different from the discrete circuit fabrication procedure. The discrete circuit starts with individual components such as resistors, transistors, diodes, etc. These components are inserted into printed circuit (pc) boards and soldered. The components are interconnected by the etched copper strips on the board. For complex circuits involving several pc boards, a mechanical assembly or rack is used to hold the pc boards. The boards are then interconnected with discrete wires.

The IC is the result of a *multistep photochemical manufacturing process.* Although the process is complex and as a whole costly, the net result is miniature devices that are inexpensive and reliable. The cost per IC is small because from one hundred to several thousand integrated circuits are made at one time. This "batch processing" type of production and the device's miniaturization are the result of intricate photography and masking. The creation of the device is physical and chemical in nature, and the procedure has evolved to a preciseness whereby device parameters are repeatable and highly acceptable by today's standards.

The study of the IC manufacturing processes represents a significant departure from customary electronic topics. Instead of components, circuits, voltages, and currents, the primary emphasis shifts to physical and chemical *processes.* The conceptual understanding of these processes, without a reasonably comprehensive background in science, may be a problem for the electronics student. However, the information is so fundamental to the understanding of integrated circuits that it cannot be avoided. The information is also essential to an appreciation of the impact of electronic products in the marketplace and their *full* potential.

2.2. SYNOPSIS

Materials are divided, electrically, into three classes: conductors, insulators, and semiconductors. The electrical properties of the semiconductor class is in between that of the conductor and the insulator, and members of this class have atomic structures that are amenable to alteration. Integrated circuits are made from semiconductor materials.

If the atomic structure of a semiconductor material is modified to produce an excess of negatively charged mobile particles (conduction band electrons), then the material is identified as *n*-type semiconductor. In *n* material, we talk about the flow of negatively charged carriers. If the atomic structure of a semiconductor material is modified to produce a deficiency of negatively charged mobile particles (valence band electrons), then the material is identified as *p*-type semiconductor. The movement of these deficiencies, or holes, is conceptually represented as the flow of positively charged carriers that we refer to in *p*-type semiconductors. The negatively charged carriers in *n*-type material are called *majority carriers*. Positively charged carriers also exist in the *n*-type material, but they are fewer in number, and are referred to as *minority carriers*. Similarly, the positively charged carriers in *p*-type material are called majority carriers and the negatively charged carriers are the minority.

Silicon is the basic element of the semiconductor class. It is rarely used in its natural or purified state, but is converted to either *n*- or *p*-type. The atomic structure of silicon is altered, to form either of the two types, by mixing into the silicon another element whose atomic structure in conjunction with silicon will produce the desired type. The added elements represent impurities in the silicon and are called *dopants*. An extremely small amount of dopant is required to change the conduction properties of silicon. A dopant used as an impurity in silicon to produce *n*-type material is phosphorus. A dopant used as an impurity in silicon to produce *p*-type material is boron.

N- and *p*-type materials, by themselves, are of little value. They can be used to fabricate resistors but little else. Their true value lies in the joining of the two to form a *p-n* junction. The operation of most solid state devices is dependent on the properties of one or more *p-n* junctions incorporated into their structure.

Several methods have been developed to form *p-n* junctions, but the technique employed in the manufacturing of ICs is the most popular. It is called *diffusion*. In this technique, the dopants exist as gases and the silicon as a solid. In diffusion, the dopant atoms of the gas penetrate, or permeate, and intermingle in the atomic structure of the solid silicon. This procedure takes place in an environment where temperature and time are of the utmost importance. The principle of diffusion is not uncommon to us. A piece of dry wood, left outside overnight, is saturated with water in the morning because of the diffusion of the water molecules from the water vapor of the dew into the wood. Ink dropped on a piece of paper diffuses through the paper structure, with the blot increasing in size as the two elements intermingle.

Diffusions can be made into other diffusions. A solid state *p-n* junction is the diffusion of a *p*-type dopant into a previously diffused *n*-type region, or vice versa. A solid state *npn* transistor is *n* diffused into *p* diffused into *n*. (In practice, the procedure is more complicated but the principle is correct.)

The IC fabrication process begins with raw silicon. Raw silicon is purified and then converted to either *p*- or *n*-type. At this stage it exists as a cylinder, called an *ingot*. From the ingot, thin circular discs called *wafers* are cut. The wafer is the foundation for hundreds of identical ICs where multiple diffusions are made. Diffusions made into other diffusions must be made with precision. There must be no overlap of the second diffusion over the first, and the second must be shallower. In fact, all dimensions of one with respect to the other must be accurately controlled. This precision must carry over to every device, since all are made simultaneously.

Diffusions are made into selected areas or locations of the wafer. This selectivity depends upon oxidation, oxide removal, and masking. The three combined simulate the use of a template in which holes are made where the diffusions can occur and elsewhere are prevented. Oxidation is the chemical process where oxygen combines with another element to form a new compound called an *oxide*. When water vapor, which contains oxygen, is passed over silicon in a high-temperature environment, a layer of silicon dioxide (SiO_2) is formed at the silicon's surface. Silicon dioxide is a high-grade insulator and is impervious to the diffusion process. It is the material that masks out the area of the wafer that is *not* to be diffused. Normally an oxide layer is grown over the entire wafer. Holes are then cut or etched in the silicon dioxide for those areas that are to be diffused.

The selective removal of the silicon dioxide is carried out by a photolithographic process using photoresistant material. In this process, a photoresistant lacquer is applied to the wafer surface. Then the photoresistant lacquer is exposed to an ultraviolet light, with a photomask being used as a template for the diffused and nondiffused areas. The ultraviolet light polymerizes the exposed photoresistant lacquer. The unexposed photoresist along with the selected oxide layer area is then removed with a solvent. The result is an opening through the oxide layer in which diffusion can take place. The wafer, covered with an oxide layer with holes in it, is placed in a high-temperature oven. A gas containing the dopant is passed over the wafer, and the impurity atoms of the gas diffuse into the unprotected areas of the silicon.

For subsequent diffusions, the process is basically repeated. A new oxide layer is grown over the entire wafer and photoresistant lacquer is applied to the surface. The lacquer is exposed to ultraviolet light, but through a different photomask. The exposed photoresist is polymerized or hardened, and the unexposed photoresist and silicon dioxide are removed with a solvent. A new area, or window, is now opened for diffusion. This procedure is repeated about six times to make an integrated circuit.

The photomasks are glass templates made to fit over the wafer. Each wafer contains hundreds of ICs, and each mask contains hundreds of patterns. Each pattern has the same dimensions as the IC. They are not made directly. The masks are initially laid out many times larger than the IC to obtain the required accuracy. They are then photographically reduced, in several steps, to the IC dimensions.

All of the integrated components are made by using diffusion, oxidation, oxide removal, and microphotographic masks. Once made, the components must then be interconnected. In printed circuit boards, interconnections are made with etched copper strips. An analogous procedure is used in ICs. It is called *metallization through deposition*. In this process, aluminum is evaporated onto the entire surface of the wafer and a photoresist sequence is repeated with a "reverse" contact photomask. The aluminum is removed from all areas except for the interconnecting traces and component contact pads.

The batch fabrication of the ICs is now complete. The wafer is cut up into individual devices, and each device is mounted and bonded to the frame of its own package. The IC device (now called a *die*) is wired by means of hairlike gold wires to the package's terminals, sealed, and then encapsulated. The device is then electrically tested, marked, packaged, and stored for distribution.

2.3. PLANAR PROCESS

The basis of solid state technology is the "planar process." It derives its name from the fact that all leads are brought out to the top surface of *plane* of the wafer, where they are interconnected or bonded. Four key concepts form the basis for this process: (a) oxidation, (b) selective dioxide removal, (c) diffusion, and (d) epitaxy. Oxidation, the formation of the silicon dioxide layer, is essential to diffusion masking, for sealing junctions, for making dielectrics for capacitors, and for acting as insulating layers for the metal interconnections. Photographic masks, an organic substance called *photoresist*, ultraviolet light, and an acidic etchant enable predetermined areas of the silicon dioxide to be etched to form the windows for diffusion. During diffusion, dopant in vapor form is deposited through the windows in the dioxide to form on the silicon's surface. Time and temperature cause these dopants to move into the selected region. Epitaxy is the *growing* of one semiconductor upon another. Diffusion or epitaxial techniques are used to fabricate *p-n* junctions.

The IC structures formed by multiple diffusions are interconnected by the condensation of aluminum vapor on the wafer surface in a process called *deposition.*

2.4. SEMICONDUCTOR MATERIAL

The foundation of the IC is a slice of semiconductor material. It is the basement of the device to which layers of other related materials will be chemically added. The electrical performance of semiconductors is between that of the conductors, such as copper and aluminum, and insulators, of which glass and quartz are examples. The most important feature of the semiconductor class of materials is that its structure and electrical properties can be modified through chemical techniques.

Although there are several members in the semiconductor class of materials, two have dominated in the production of semiconductor-related components. Germanium is widely used in the manufacturing of transistors. However, today, all integrated circuits and many transistors are made with silicon. It is chemically simpler than most other semiconductors and has very good high-temperature electrical properties. This is important in the military and aerospace applications of ICs, where operating environments reach temperature highs in excess of $100°C$ and temperature lows approaching $-50°C$. However, the prime attribute of silicon is its ability to grow a stable oxide. This means that the silicon can combine, at its surface, with gaseous oxygen to form a layer of insulating material called silicon dioxide (SiO_2). The process is called *oxidation.* Silicon dioxide, the product of oxidation, plays a key role in the IC fabrication. It is one of the layers that is chemically built on the slice of silicon. The time and extent to which silicon has been used in the production of ICs has made it a highly developed industrial technology.

There are two main requirements in preparing the silicon for solid state devices. First, the silicon must be extremely pure. Unwanted impurities must be down to a level of 1 part in 10^{10}, i.e., one impurity atom for 10,000,000,000 atoms of silicon. Second, the silicon must have a continuous and regular crystal structure. Silicon atoms have an irregular structure. The orientation of the atoms must be aligned into what is called *single-crystal form.*

The purification of silicon is performed through the chemical processes of distillation and reduction. The result of these processes is the depositing of silicon onto the surface of a high-purity silicon rod, building it up to a diameter between 1 and 4 inches. The silicon deposits on the rod are in polycrystalline form; that is, their atomic structures are joined together in random orientation.

The procedure generally used to produce single-crystal silicon for ICs is called *crystal pulling*. Purified, solid silicon is placed in a crucible within a furnace and heated to a temperature just above its melting point. A seed crystal, a small piece of single-crystal silicon, is lowered until it barely enters the melt or molten silicon. Because the seed crystal is at a lower temperature than the melt, heat flows from the melt to the seed. The temperature of the melt in contact with the seed falls, and some of the melt solidifies onto the seed, with the atoms arranging themselves to have the same orientation as the atoms in the seed crystal. The seed crystal is rotated and slowly raised, growing larger as more silicon solidifies onto it. Typically, pulled silicon crystals are cylindrical in shape, between 2 and 4 inches in diameter, and 8 to 40 inches long; they are called *ingots*.

An impurity dopant, *p* or *n*, is added in the initial melting process so that the pulled crystal has the required conduction properties.

In the fabrication of transistors and ICs, the silicon is used in the form of thin circular slices. The *p* or *n* ingots are sawed into slices about 20 mils thick by a thin diamond-impregnated wheel rotating at high speed. The slices are mechanically lapped to smooth the surface. Finer and finer polishing compounds, or chemical etchants, are used until the surface is flat and mirrorlike. The slices of silicon, called *wafers*, pure and single-crystal in structure, are then ready for subsequent processing. Ultimately from each wafer will come one-hundred to several thousand ICs. Each IC on the wafer is referred to as a die (the plural is dice). Depending on the complexity of the IC, die sizes vary from 0.03 to 0.250 inch square. Figure 2-1 highlights the steps in the preparation of silicon for IC manufacturing, and Figure 2-2 is the resultant product.

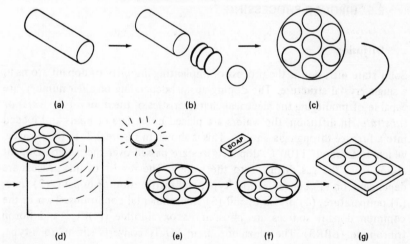

(a) (b) (c)

(d) (e) (f) (g)

Figure 2-1. Preparation of the IC Wafer. (a) The basic ingot; (b) slicing the wafer; (c) mounting the wafer in wax on a plate; (d) lapping; (e) polishing; (f) cleansing; (g) storage and handling.

Figure 2-2. A Silicon Ingot and Wafers

2.5. BIPOLAR PROCESSING

Diffusion

Solid state diffusion is the process of implanting impurity or dopant atoms in a single-crystal structure. The doping atoms, depending on their number, are capable of modifying the electrical characteristics of the *n*- or *p*-type *wafer or substrate*. In diffusion, the wafers are placed in carriers or boats and loaded into a furnace through glass tubes. This is shown in Figure 2-3. In the furnace, which is at about 1100°C, dopant gases are passed over the wafers. Dopant atoms from the gas diffuse into the wafers. The results of this process are dependent on (1) concentration, type, and distribution of dopant, (2) time, (3) temperature, (4) pressure, and (5) environmental conditions. Two of the common dopant sources are phosphorus oxychloride (POCl$_3$) and boron tribromide (BBR3). The phosphorous in POCl$_3$ converts silicon to *n*-type, and the boron in BBR3 converts silicon to *p*-type.

Several diffusions are made in fabricating ICs. Each one is labeled with a name related to the function or component lead with which the diffusion is

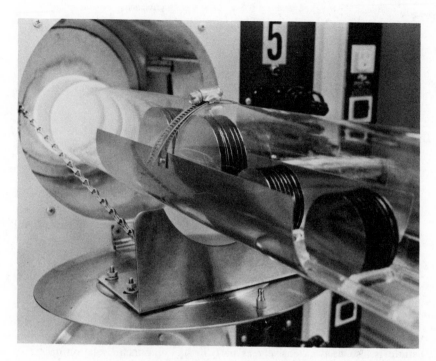

Figure 2-3. Wafers Loaded into a Diffusion Furnace

associated. The isolation diffusion creates separate regions in the epitaxial *n*-layer (on a *p* wafer) for each IC component. The epitaxial *n*-layer is a non-diffused layer and forms the collectors of *npn* transistors, and the cathodes of circuit and isolation diodes. The isolation diffusion electrically separates the various areas of the wafer containing circuit elements. The base diffusion forms the base of *npn* transistors, and the anodes of circuit diodes and *p*-type resistors. It is a shallow diffusion requiring closer process control. The emitter diffusion forms the transistors' emitters, low-value *n* resistors, MOS capacitor plates, and low-resistance terminals for *n* collector regions. A fourth diffusion called a *buried layer* is sometimes used and is discussed under IC components. Its function is to form a low-resistance region under the collectors of *npn* transistors. The four key diffusions are illustrated in Figure 2-4.

Epitaxy

A second, often used, technique of fabricating a *p-n* junction is called *epitaxy*. Epitaxy is the growing or deposition of one semiconductor upon another. It is a growing process in which gaseous molecules collect in regularly oriented patterns on the outside surface of a solid material. The process is electrically

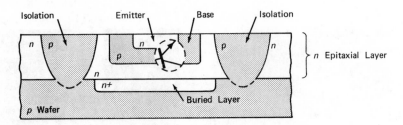

Figure 2-4. Key Diffusions in the IC Fabrication Process

and thermally induced. Epitaxy differs from diffusion in that the epitaxial interface is grown, whereas the diffused interface is the result of a molecular formation. An epitaxial grown layer is more easily process-controlled and possesses certain advantageous electrical properties, but is more expensive.

An epitaxial layer is usually grown over a *p* wafer as one of the initial steps in the formation of integrated *p-n* structures. This layer takes on the same crystal orientation as the single crystal wafer. *P* isolation diffusions split the epi layer, Figure 2-5, into sections or wells for the individual components. The epitaxial *n* layer is the collector of *npn* transistors. The *p-n* junction formed by the epitaxial grown layer of *n*-type material upon a *p*-type substrate is also used extensively as part of the component isolation necessary for ICs.

Oxidation

Silicon chemically reacts with the oxygen of water vapor to form an oxide called silicon dioxide (SiO_2). Basically, time and temperature determine the thickness of the oxide on the wafer surface. The properties of silicon dioxide make it one of the most useful of microcircuit materials. The formation of

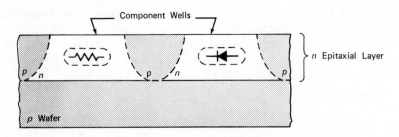

Figure 2-5. Epitaxial Layer Split into Component Wells

the oxide layer is easily controlled, the layer is uniform and continuous, and it closely adheres to the silicon surface. It is also an effective barrier for the more important diffusants but is easily removed by using a chemical solvent. These two factors make SiO_2 ideal as a diffusion template or screen. The oxide layer is a good dielectric, finding further value in the fabrication of capacitors and passivating or protecting surfaces.

Oxide Removal

The photochemical procedure of removing selected areas in the oxide layer is similar to that used in the manufacturing of printed circuit boards. The wafer with the oxide layer is uniformly coated with a layer of a chemical called *photoresist*. The photoresist is dried and covered with a glass photomask that contains the desired patterns. The photoresist is then exposed to ultra-violet light through the mask. The nonopaque areas are polymerized, i.e., converted to a new compound, which resists attack by acids and solvents. The nonpolymerized photoresist and oxide are etched away with an acidic solvent. The wafer is now ready for diffusion, which will occur only in the openings, called *windows*, in the oxide. The complete photoresist process is repeated each time the silicon dioxide is selectively removed. The selective oxide removal by the photoresist process is illustrated in the step-by-step procedure of Figure 2-6.

Masking

The templates or screens used to define the regions for diffusions and metallization are called *masks*. The fabrication of an IC requires from five to as many as twenty different masks, with six a typical number. A six-mask set is shown in Figure 2-7. The masks are used one at a time, during each of the processing steps. Each one consists of a unique pattern established by a photographic emulsion on a flat glass plate. Areas of the photomask are either transparent or opaque to light. The masks are the same size as the wafer and must be highly accurate to define the proper regions for processing. This high degree of accuracy and resolution is achieved through a photographic process. The patterns are initally drawn, a Mylar base and tape being used, 200 to 1000 times the actual size; they are called, at this stage, the artwork. The artwork is then reduced, in several steps, to the IC dimensions by using micro-photographic techniques. The images are transferred to a glass disc, which is now called a *photomask* (Figure 2-8). Since a wafer contains many ICs, the pattern is photographically duplicated and repeated many times to complete the wafer mask.

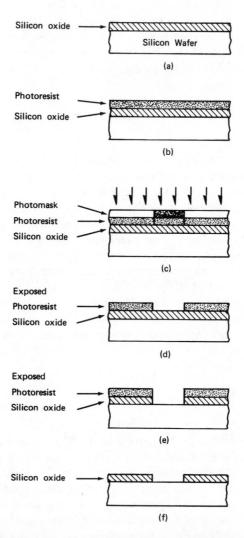

Figure 2-6. Selective Oxide Removal by the Photoresist Process (a) Silicon wafer with oxide formed on the surface; (b) photoresist lacquer applied to the surface; (c) photoresist exposed to ultraviolet light through a photomask; (d) unexposed photoresist removed with solvent; (e) silicon oxide removed by etching; (f) photoresist removed to leave window in silicon oxide.

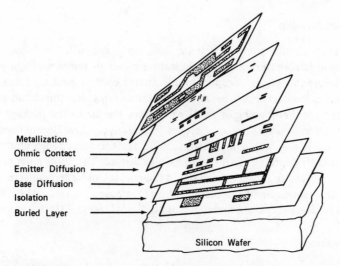

Metallization
Ohmic Contact
Emitter Diffusion
Base Diffusion
Isolation
Buried Layer

Silicon Wafer

Figure 2-7. Photomask Set Used to Fabricate IC's. Buried layer mask—used to form a low-resistance N+ region under the active devices (optional). Isolation diffusion mask—used to create separate N regions or component wells for the IC components. Base diffusion mask—used to form the base region of transistors, one side of a diode, and resistor areas. Emitter diffusion mask—used to form the transistor's emitter, low-value resistors, and one side of a capacitor plate. Ohmic contact mask—used to make holes through the oxide layer to permit the metallization to make contact to the terminal pads of the circuit elements. Metallization mask—used to form the metal interconnects between the circuit elements and to provide external bonding pads.

Figure 2-8. Photomask

Metallization

Integrated circuit components are interconnected with metallic, usually aluminum, traces during the metallization phase of the fabrication process. Also during this phase, component and circuit contact pads are formed. The component contacts must be ohmic or nonrectifying, and the circuit pads are provided for the bonding of small wires from the die to the package's leads. Metallization occurs through a process called *deposition*. During deposition, aluminum is evaporated in a high-vacuum system, and enough aluminum is permitted to deposit over the entire wafer surface. The photoresist process is again employed to protect the metal in the contact and interconnection areas, while the remainder of the metallization is exposed for removal by etching. The mask and process are inverted from that used in the diffusion steps.

Probing

The last step in wafer processing is to test the die. At this stage, a machine called a wafer prober makes contact to the die through minute, pointed probes. These probes touch the aluminum pads of the IC and provide the electrical connection from a test instrument or system to the IC. Each die is electrically tested against predetermined specifications. Those dice that are determined to be faulty are inked and rejected. Figures 2-9 and 2-10 show two perspectives of the wafer probe procedure. Figure 2-9 is a photomicrograph of the prober needles on the circuit pads of an IC. Figure 2-10 is a macroscopic view of the probe heads with their needles making contact with a wafer die. The prober automatically indexes from die to die, resting only to perform the electrical tests and inking, if necessary.

Dicing

After the dice have been tested, they are ready to be separated by using a procedure called *dicing*. The most common method is scribing and breaking. Fine, diamond-cut lines are scribed vertically and horizontally across the wafer but between each circuit or die. The wafer is then placed on a rubber pad, and pressure from a roller is exerted on the wafer, breaking it up into individual chips.

Assembly

Assembly is the most expensive part of the fabrication process. In wafer processing, large numbers of circuits are processed simultaneously, but during assembly the dice are handled individually, and connections to each die are made separately. Assembly operations are required to protect the die, to facilitate handling, and to provide a package and contacts for equipment application.

Figure 2-9. Prober Needles on an IC Die

Figure 2-10. Prober with Needles Testing a Die on a Wafer

Die Bonding. Die bonding is the soldering, brazing, or glassing of the die to a package frame. This attachment provides a mechanical and, occasionally, an electrical contact from die to package. This bond also acts as a thermal path for internally generated heat to flow to the surroundings.

Wire Bonding. The electrical connections from the chip to the package's terminals are made by one of several wire bonding techniques. One-mil (0.001 inch) wire, either gold or aluminum, is connected to an IC aluminum pad and then extended to the package terminal, where a similar connection is made. Ultrasonic, pressure, or temperature techniques are used to form the wire to pad and terminal welds. Great emphasis is being placed on eliminating this costly and labor-oriented method and replacing it with an automated procedure. Presently most IC manufacturers have shifted to an automated bonding technique where connections are made from die to the package by using metal tabs.

Packaging. The individual IC die is too small and delicate to handle. To facilitate handling and protect it from damage, the die is welded to a frame, its leads are connected, and the unit is sealed, passivated, and encapsulated. Glass, ceramic, and plastic are common materials that encase the die. There are many types of IC packages, but the most popular are the axial lead (TO5), flat pack, and dual-inline (DIP) packages. They are illustrated in Figure 2.11.

Testing

The last step in the manufacturing process is the electrical testing of the IC. DC, AC, and functional measurements are made to insure that the IC will perform to established standards. Large-volume products are automatically transported by mechanical autohandlers. A computer-controlled IC test system, in conjunction with the autohandler, systematically and quickly screens each IC to verify its performance. Bad ICs are rejected, and those accepted are marked and stored for distribution. Figure 2-12 shows a modern facility for testing ICs.

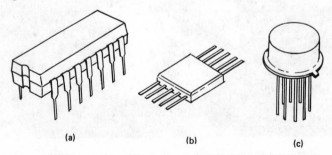

(a)   (b)   (c)

Figure 2-11. Common IC Packages. (a) Dual inline (DIP); (b) flat pack; (c) TO-5.

Figure 2-12. IC Test Facility

2.6. UNIPOLAR (MOS) PROCESSING

The manufacturing, or processing, of MOS integrated circuits uses essentially the same technology required for the bipolar ICs. Doped silicon wafers are used for the foundation; each device has a unique set of masks; silicon dioxide is deposited for use as an insulator; windows are etched in the dioxide layer, and diffusions are made. The components are interconnected with aluminum traces; the dice are tested, separated, and bonded to their package's leads. The packages are then sealed and retested, and are ready for shipment.

MOS transistors are simpler to fabricate than their bipolar counterparts and require a smaller number of processing steps. Since MOS transistor action takes place at or very near the device surface, the control of the diffusion step is less critical. In addition, the self-isolating features of MOS structures do not require special isolation provisions required by the bipolar structures in integrated circuit form. These MOS advantages, however, are offset by severe requirements for surface cleanliness, gate pattern accuracy, and oxide thickness accuracy.

The sequence of monolithic IC process steps is illustrated in Figure 2-13. Although how they make the MOS IC is similar to the bipolar IC, what they make differs. These differences are highlighted when the MOS and bipolar component structures are examined.

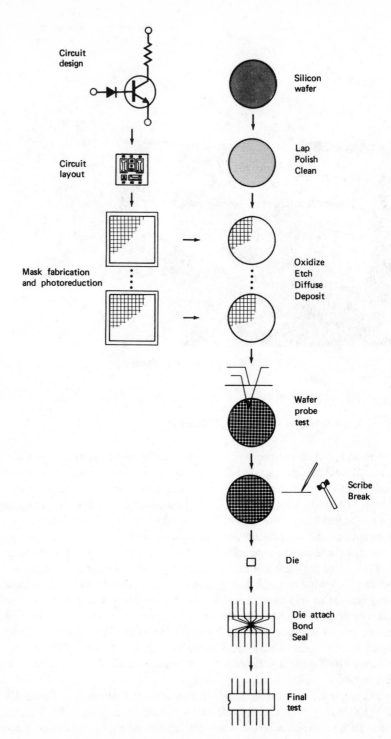

Figure 2-13. Monolithic IC Processing Steps

24

QUESTIONS

1. What manufacturing characteristic makes the IC inexpensive?
2. List one *n*- and one *p*-type dopant. In what form is the dopant when it is diffused into the silicon? From what compounds are these *n* and *p* dopants derived?
3. To what degree is raw silicon purified before it is used in fabricating ICs? What is unique about the structure of the atoms in the purified silicon wafer?
4. Why is silicon preferred over germanium in the manufacturing of integrated circuits?
5. Define the four key process concepts that form the basis of present IC production technology. For each case, list one associated material.
6. What numbers are associated with
    a. Wafer thickness?
    b. Minimum die size?
    c. Maximum die size?
    d. Wafer diameter?
7. What factors affect the diffusion process?
8. Identify the differences between an epitaxial layer and a diffused layer. What are the advantages and disadvantages of each?
9. List three functions of the silicon oxide layer. What determines how thick the oxide layer is on the wafer surface?
10. In which facet of the production of an IC does microphotography play a role? Why? If the diameter of a wafer is 3 inches and it contains 500 dice, estimate the size of the photomask's artwork.
11. What is a basic difference between a metallization photomask and a diffusion photomask?
12. What two basic functions does metallization accomplish? What metal is usually evaporated in this process?
13. Why is assembly the most expensive part of the fabrication process? What aspect(s) of the assembly process is currently automated?
14. How are connections from the die to the package's leads made?
15. What is the minimum number of times that an IC is electrically tested? When is it tested?

3

Integrated Circuit Components

Before one can understand the whole, he must know the parts. So it is in integrated circuits. This chapter discusses the parts or components that make up the integrated circuit.

3.1. INTRODUCTION

The designer of discrete circuits has a large number and wide variety of components that he can use. He can select from resistors, inductors, and capacitors to transformers, switches, and indicators. He can specify accuracy, voltage, current, and power ratings, size, and even shape. The designer of monolithic circuits is not so fortunate. He has a restricted number of components that he can use, and their performance and ratings is process-limited. However, he is not totally at a disadvantage. He is less restricted by space and can use a greater number of components. Most important, he can produce his circuit in large quantity for significantly less cost.

The components that may be integrated are resistors, capacitors, diodes, and transistors. Capacitors are costly to integrate and are used sparingly. Inductors cannot be integrated, but their effect can be simulated through circuit techniques. In fact, the circuit design techniques of solid state technology have allowed designers to duplicate most electronic functions.

BIPOLAR COMPONENTS

3.2. DIODES

A solid state diode is formed by making an *n* diffusion into a *p* wafer (or vice versa). The *n* area is the diode's cathode, and the *p* area is the anode. Diode action occurs at the *p-n* junction. When the positive terminal of a voltage supply is connected to the anode and the negative terminal is connected to the cathode (Figure 3-1) the diode is forward-biased. The positive voltage attracts the negatively charged carriers of the *n* region and they flow through the *p-n* junction. The carriers flowing from the *n* region are replenished from the source's negative terminal. Similarly, the negative voltage attracts the positively charged carriers of the *p* region. They flow through the *p-n* region to the source's negative terminal, resulting in the flow of current. As the voltage is increased, the current rapidly increases.

When the positive and negative terminals of the supply are connected to the cathode and anode, respectively, the diode is reverse-biased. The negative voltage on the anode attracts the positive carriers of the *p* region but repels the negative carriers of the *n* region. A similar action occurs at the cathode. The repelling of the positive and negative carriers in the diode leaves the *p-n* junction void of any carriers. No current flows, and the diode appears as an open circuit.

Circuit and Isolation Diodes

A diode is a *p-n* junction. The *p-n* junctions that function as diodes in integrated circuits are used for two purposes: (a) circuit action and (b) isolation.

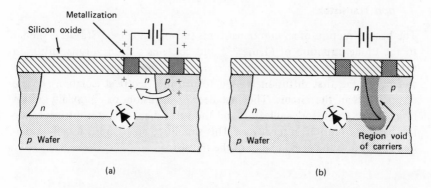

(a)   (b)

Figure 3-1. A p-n Junction. (a) Forward-biased; (b) reverse-biased.

Figure 3-2 shows two integrated diodes. They are formed as the result of two diffusions. The *p-n* junction formed by the second diffusion is the *circuit diode*. The first *p-n* junction, formed from the *p* wafer and *n* diffusion[1], is an *isolation diode*. The *n* region may be diffused or may be an epitaxially grown layer. Isolation diodes are used to isolate integrated circuit components from each other. To provide isolation, a contact is made to the *p* wafer and is externally attached to the most negative voltage supply used by the IC. Since this negative voltage is on the anodes, the isolation diodes are reverse-biased and appear as open circuits. Each circuit component sees an open or reverse-biased diode to the wafer, and hence to each other. Thus IC components can operate independently and with minimal interaction.

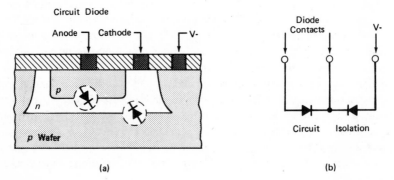

(a)   (b)

Figure 3-2. IC Diodes; Circuit and Isolation. (a) Integrated structure; (b) equivalent circuit.

3.3. TRANSISTORS

npn Transistor

The basic structure of an *npn* transistor is shown in Figure 3-3(a). It is similar to the diode structure of Figure 3-2, but contains a third, *n*-type, diffusion made into the second diffusion's *p* region. The *p-n* junction formed by the wafer and the first diffusion is used to provide electrical isolation for the components in the circuit. The first deep *n* diffusion, or epitaxially grown layer, is the transistor's collector, and the subsequent *p* and *n* diffusions are the base and emitter, respectively. This simplified structure is usually modified to improve the performance of the transistor.

---

[1] For simplicity in this chapter, this *n* region is generally implied to be diffused. In most ICs, it is epitaxially grown.

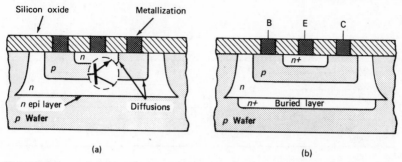

Figure 3-3. Integrated Structure of a Bipolar (npn) Transistor. (a) Simplified; (b) with *n+* diffusions.

Minimizing Resistance of Diffused Regions and Component Contacts in IC Transistors

The resistance of the collector region of an *npn* transistor adversely affects the performance of the device in amplification and switching circuits. This resistance can be reduced by shunting it with a smaller value. In an IC, this is accomplished through an additional diffusion of *n* material that has an extremely high concentration of dopant. This diffusion, called n^+, is located below the collector region and is referred to as a buried layer. The resistance of the n^+ layer is much smaller than the collector because of the greater amount of carriers in the dopant. Some n^+ and p^+ diffusions are also used to minimize the contact resistance between the die and the interconnection metallization.

Figure 3-3(b) shows an IC structure for an *npn* transistor including the n^+ buried layer and n^+ emitter diffusions. The shallow emitter diffusion is usually entirely n^+.

Windows are cut in the oxide layer for deposited metal to interconnect the leads of the transistor to the circuit. The mask associated with this step is called the *ohmic contact mask.*

pnp Transistor

The bipolar *npn* transistor is the most widely used three-terminal device. Its overall performance is superior to that of its complement, the *pnp* transistor, and is less difficult to manufacture. However, the *pnp* transistor is important in many applications.

The structures of the *npn* and *pnp* are shown in Figure 3-4. The *pnp* requires an additional p^+ diffusion step for the device's emitter, and is referred to as a *complementary pnp*, because its structure is analogous to the *npn*.

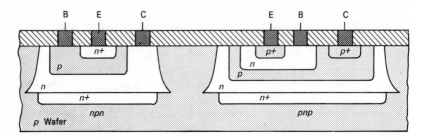

Figure 3-4. Integrated Structures for the npn and Complementary pnp Bipolar Transistors

Two additional methods of fabricating *pnp* transistors are used: (1) the lateral *pnp* and (2) the vertical *pnp*. They are shown in Figure 3-5.

The lateral *pnp* starts with two *p* regions diffused into an isolated *n*-type island and is made at the same time as the base diffusion (*p*) for the *npn* transistor. Two *p* diffusions, both circular, are made into the isolated *n* region for the device's collector and emitter. The current flows laterally from the emitter through the two base regions to the collector. This device is used extensively in linear ICs.

The vertical *pnp* uses the *p* substrate for its collector. Since the substrate is connected to the most negative IC voltage, it is restricted in the IC design to those *pnp* applications that tie the collector to V^-. The emitter is a *p* diffusion into the *n* base. Current for this device flows from emitter to base and vertically to the collector.

Whether the bipolar transistor is *npn* or *pnp*, it fundamentally depends on current for its operation. When positively charged carriers (*npn*) or negatively charged carriers (*pnp*) are injected into the transistor's base lead, a resultant gain in current is present in the device's emitter and collector leads. The mode of operation of the current-controlled bipolar transistor is significantly different from the voltage-controlled field–effect (MOS) transistor.

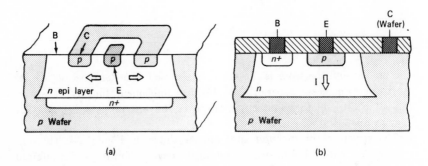

Figure 3-5. Integrated Structures for the Lateral and Vertical pnp

Operating Points

The principal component in all ICs is the transistor. Unlike the passive components, i.e., resistors, capacitors, and diodes, it is capable of gain and it is called an active device. Although most circuits take advantage of the important gain characteristic, the transistor is not exclusively restricted to amplification applications. It is also used as a switching device and as a simulated passive load. Its operation in digital and linear circuits differs, and in fact, can vary within a given IC. The range of operation of a transistor in a circuit is represented by a load line drawn on the device's (bipolar) collector family of curves. These curves are a two-dimensional graph of collector current (I_c) versus collector to emitter voltage (V_{CE}) for various values of I_B. Figure 3-6 shows a collector family of curves for an *npn* transistor. This graph, with the load line, illustrates the modes of operation, generally, of transistors in digital and linear integrated circuits. Of course, similar graphs exist for unipolar (MOS) transistors. In digital ICs, the circuits are designed so that the transistor operates *at cutoff* or *at saturation*. In linear ICs the circuits are designed such that the transistor operates *between cutoff* and *saturation*. In digital ICs, the transistor is operated as a switch that is ON at its saturation point and OFF at its cutoff point. In linear ICs, the transistor is operated in the area between these two points, which is referred to as the *linear region*. The static or start point of the linear region is called the quiescent or Q point. Indirectly, digital circuits take advantage of the transistor's gain capability. However, linear circuits directly depend on it.

In digital circuits, speed and power consumption are of the utmost importance. In linear circuits, gain and power consumption are the most important device criteria. For each of these circuits, the transistors are specifically designed to maximize these characteristics. This is accomplished by varying the geometry and content of the transistor's IC structure.

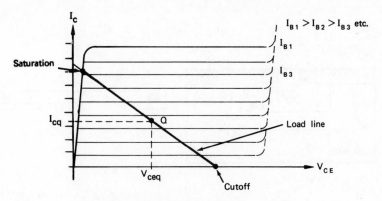

Figure 3-6. Bipolar Transistor Family of Curves

3.4. RESISTORS

Silicon is a resistive material. In fact, all materials have resistance. For insulators such as glass, mica, or silicon dioxide the resistance is very high but not infinite. For conductors such as copper, gold, or silver, the resistance is relatively low but not zero. We approximate a short length of copper wire as a short circuit or 0Ω, but our approximation would not hold true for a mile of wire. In fact, the resistance of a mile of wire is several hundred ohms (24 AWG) and increases as we increase the length of the wire. The value of resistance of a piece of wire not only depends on its length but also is a function of the diameter and the type of material. The characteristic that relates length, diameter, and material to resistance is called *resistivity*. The concept of resistivity applies to all materials, insulators, conductors, and semiconductors.

Figure 3-7 shows the cross section and top view of a *p*-type monolithic resistor. The *p-n* junction of the *n* epi layer and *p* wafer forms the isolation diode. A *p* diffusion is then made into the *n* area. The monolithic *p* resistor that exists is continuous; i.e., its resistance increases as we leave one terminal and move toward the other. Its final or maximum value is a function of the *p* diffusion's thickness, its length, width, the type of material, or collectively, the material's resistivity. Unlike the pure copper wire, the diffused *p*-type material can be altered to change the resistance. The alteration is in the form of controlling the concentration and type of dopant or impurity of the *p* material. Thus dopant concentration and type control the resistivity of the semiconductor material, and this, with length, width, and thickness of the diffused material, controls the resistor's value.

There are several types of IC resistors. The previously mentioned resistor is called a *diffused-base type*, because it is formed simultaneously with the transistor base. Similarly, *diffused emitter* resistors are formed simultaneously

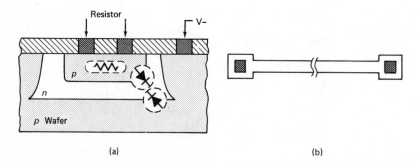

(a)   (b)

Figure 3-7. IC Resistor: Diffused Base Type. (a) Integrated structure; (b) top view.

with the transistor emitters. These resistors are useful where low values are required. To achieve high resistance values, *base pinch resistors* are employed.

Large values of resistance may be obtained with minimum die area by squeezing or pinching the base region. The *p* base region is pinched by the emitter diffusion (see Figure 3-8), which results in a greater resistance per length. The smaller the *p* region is made, the fewer carriers it will have, and hence the greater its resistance. No contact is made to the emitter diffused area, and it serves only to increase the *p* region resistance. The nominal pinched based resistance is 10 kΩ per square mil (0.001 in. $\times$ 0.001 in.).

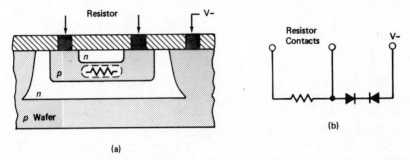

(a)

(b)

Figure 3-8. IC Resistor; Pinched Base Type. (a) Integrated structure; (b) equivalent circuit.

Diffused base resistors have a fairly good temperature coefficient, but, as is true with any IC resistor, they have poor absolute accuracy. This is somewhat compensated for by the excellent relative accuracy and tracking ability they have in conjunction with other device resistors. A typical resistor requires at least 1 square mil (0.001 in. $\times$ 0.001 in.) per 100 Ω. For a width of 1 mil, the resistor would have to be 20 mils long for a value of 2 kΩ. Large-valued base diffused resistors can increase the die size and the cost through a loss in yield and materials.

3.5. CAPACITORS

Diodes, resistors, and transistors are the fundamental bipolar integrated components; however, a limited range of low-value capacitors is fabricated and used. A reverse-biased *p-n* junction, while theoretically an open, is in actuality a small capacitor. Capacitor action occurs when a change in voltage causes a corresponding change in stored charge or carriers. To understand how a diode behaves as a capacitor, we examine its structure under the reverse-biased condition as shown in Figure 3-9. As is true in any IC, the first diffusion is used to isolate the IC components.

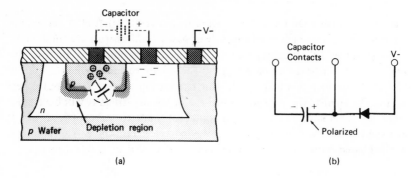

Figure 3-9. IC Capacitor: Junction Type. (a) Integrated structure; (b) equivalent circuit.

The second p and n diffused regions shown in the figure have an excess of positive and negative carriers, respectively. This is the result of the introduction of the impurity dopants used in the diffusion process. When a positive voltage is applied to the diode's cathode (n) and a negative voltage is applied to the anode (p), these excess carriers are drawn to the voltage source. The positive carriers are drawn to the –voltage, and the negative carriers are drawn to the +voltage, depleting the region of carriers at the p-n junction. This depletion region is like an insulator. That is, it has no carriers, no current flow, and an extremely high resistance. The magnitude of voltage influences the amount of stored charge. As the voltage increases and decreases, a proportional amount of charge is increased and decreased. Therefore, the junction behavior exhibits the same characteristics as a parallel plate capacitor. Unlike the discrete capacitor, the value of the bipolar junction capacitor is voltage-dependent since the applied voltage also affects the depletion region width. Since the more positive voltage must always be applied to the cathode, the junction capacitor is polarized. The value of capacitance per diffused area is low, and economic considerations restrict the size and usage. A typical value of an integrated bipolar capacitor is 30 pF, with a maximum of about 100 pF.

3.6. PHYSICAL CHARACTERISTICS

Component Profiles

Figure 3-10 shows a side section and top view of the four bipolar components, including the interconnection metallization. It should be noted that the n and p diffusions for *all* components of *all* dice are done simultaneously and that their integrated structures are similar. The geometries of the diffusions will vary, depending on the desired electrical performance.

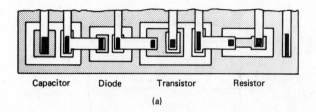

| Capacitor | Diode | Transistor | Resistor |

(a)

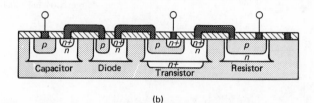

(b)

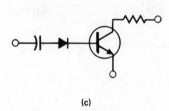

(c)

Figure 3-10. Component Profiles and Equivalent Circuit. (a) Top view; (b) side view including metallization; (c) equivalent circuit.

Component Dimensions

The approximate vertical dimensions of a component on a wafer are shown in Figure 3-11(a). This figure graphically illustrates the preciseness and miniaturization associated with the production of the IC. These dimensions are brought into proper perspective when one considers that the diameter of a thick hair is about ten mils.

Figure 3-11(b, c, and d) show the top view of the three fundamental IC components and the typical horizontal dimensions of each. Great emphasis is placed on minimizing each component's surface area to insure the smallest die size possible. Yield, i.e., the number of good dice per wafer, is directly related to the area of the integrated circuit. The greater the yield, the lower the cost.

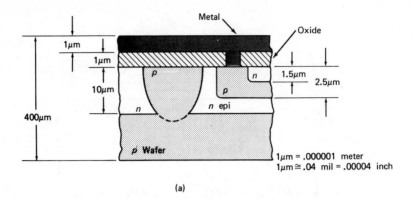

(a)

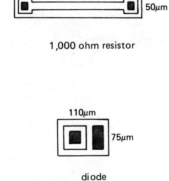

(b)

Figure 3-11. (a) Typical Vertical Dimensions of an Integrated Circuit. (b) Typical Horizontal Dimensions of IC Components

UNIPOLAR (MOS) COMPONENTS

3.7. CAPACITOR

A physical capacitor is defined as two metal plates separated by a dielectric or insulator. The MOS capacitor follows that definition and illustrates a second important role of the dioxide layer. The structure of the MOS capacitor is shown in Figure 3-12. The capacitor contacts are metallized and are separated by a thin layer of silicon dioxide, an insulator. The n^+ diffusion provides for a low-resistance contact for the capacitor's bottom lead. Since the capacitor requires a large area on the IC, its value and usage are limited. The typical value of a MOS capacitor is 30 pF.

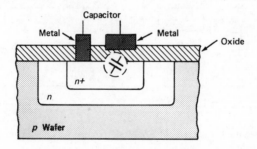

Figure 3-12. The Unipolar (MOS) Capacitor

3.8. TRANSISTORS

p-MOS Transistor

The most common MOS structure, the p-MOS transistor, is illustrated in Figure 3-13. The device is built on an n-wafer and contains two p^+ diffusions for the device terminals, called *source* and *drain*. The third terminal or metallized gate is separated from the wafer by a thin layer of silicon oxide, and its voltage (with respect to the source) is the input control variable. The source and drain are isolated from other components by the p-n junction formed with the wafer. The gate is isolated by the layer of silicon oxide.

The n area between the source and drain is called a *channel*. With a negative drain-to-source voltage, and zero volts potential difference between the gate and source, the channel is nonconducting, and no drain-to-source current flows. A negative gate-to-source voltage, Figure 3-13(b), will *induce* a positive charge at the surface of the channel just beneath the oxide layer. This constitutes a conductive path between the two *p*-type terminals, resulting in the flow of drain-to-source current.

The MOS transistor is a three-terminal device. Its input-output relationship is a function of the input voltage (V_{GS}) and the output current (I_{DS}). Because an *increasing* gate voltage causes a greater drain current to flow, the device is said to operate in the enhancement mode. A fourth lead is associated with MOS devices and is called the *substrate* or *bulk*. It is a connection to the wafer and is usually connected to the source to prevent interaction and erroneous currents between the other leads.

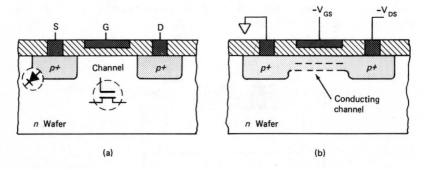

Figure 3-13. Integrated Structure of a *p*-MOS Transistor. (a) Nonconducting channel; (b) conducting channel.

n-MOS Transistor

The relationship of the *n*- and *p*-MOS transistors is similar to that of the *npn* and *pnp* bipolar transistors. A simplified *n*-MOS structure is shown in Figure 3-14. The device is built on a *p*-type wafer and contains two n^+ diffusions for the source and drain. The metallized gate is isolated from the wafer by a thin layer of silicon oxide.

The channel between the source and drain is either conducting or nonconducting. With a positive V_{DS} and zero V_{GS}, the channel is equivalent to an open and no drain-to-source current (I_{DS}) flows. The application of a positive V_{GS} will induce a channel between the source and drain, resulting in the flow of current. The device operates in the enhancement mode.

While theoretically the operation of the *p*- and *n*-MOS transistors are the same, the performance of the two varies because of physical parameters. A number of processing techniques are now used to modify the performance levels of each, of which silicon gate and ion implantation are but two examples.

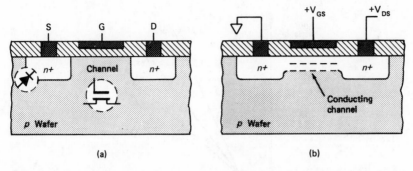

Figure 3-14. Integrated Structure of a *n*-MOS Transistor. (a) Nonconducting channel; (b) conducting channel.

Transistor Operation

The operation of MOS transistors differs from that of bipolar devices. The term *bipolar* refers to the two polarities of current that exist in devices of this type. Both holes and electrons or positively charged carriers and negatively charged carriers are essential to their operation. MOS devices are *unipolar*, and only one type of carrier is necessary in the operation of a particular MOS transistor.

The active leads of a bipolar transistor are called the *emitter*, the *base*, and the *collector*. The active leads of a unipolar transistor are called the *source*, the *gate*, and the *drain*. The active *region*, where transistor action occurs, for bipolar devices is in the base. The base is *beneath* the surface and between the emitter and collector. MOS devices are surface-effect devices. Their active region consists of a channel that is induced (enhancement mode operation) at the gate and oxide interface.

The MOS transistor is a voltage in, current out device. Figure 3-15 is a graph of the output current I_D versus the input voltage V_{GS}, for several values of V_{DS}. The device is off ($I_{DS} = 0$) until V_{GS} is greater than the turn-on or threshold voltage V_T. The slope of the transfer curves is the transconductance g_m. This parameter can be used as the gain figure or transfer function of the MOS transistor. Transconductance g_m decreases with rising junction temperature. Current gain, β or h_{FE}, in bipolar devices increases with rising temperature.

There are two modes of operation for field effect devices: enhancement and depletion. Junction FETs are exclusively depletion mode devices but MOS FETs may operate in both modes. Most MOS ICs use enhancement mode devices that are nonconducting with zero volts applied to the gate. A conducting channel is created or enhanced by an electrostatic field which is produced by a gate-to-source applied voltage. *P*-MOS, or *p* channel devices, require negative gate-to-source voltages, and *n*-MOS requires positive voltages.

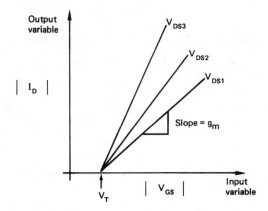

Figure 3-15. MOS Transistor Input-Output Transfer Curves

The gate of an MOS transistor is electrically isolated, by the oxide layer, from any other transistor part. As a result, the DC input resistance is extremely high.

The drain family of curves for an MOS enhancement transistor is shown in Figure 3-16. The linear region for transistor operation is defined from V_{DS} (MIN) to the breakdown voltage. The operation of the device in the nonlinear region is important in MOS ICs also because the V-I characteristic is approximately resistive.

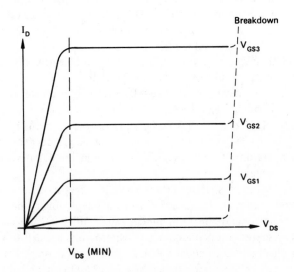

Figure 3-16. MOS Transistor Family of Curves.

3.9. RESISTOR

A significant advantage that the MOS process possesses over the bipolar process is the smaller IC area required for the MOS transistor. Normally, the MOS transistor is operated as a switch; that is, it is either on hard or off. However, in between these two states, it behaves as a voltage-controlled resistor. It is used as a resistor by setting the voltage on the gate to a fixed value that turns the transistor permanently on. This voltage and the channel's geometry and concentration will determine the resistor's value. Its value normally is many times that of the ON resistance of a switching transistor, and values in the 100 kΩ range can be achieved with minimal die area. Figure 3-17 shows an MOS device used as a drain load resistor for an MOS switching transistor.

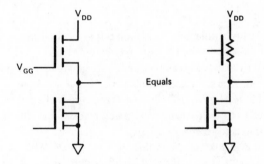

Figure 3-17. The Unipolar (MOS) Resistor

3.10. COMPLEMENTARY MOS (CMOS) TRANSISTORS

Figure 3-18 illustrates the integrated structure for p- and n-MOS transistors on the same substrate or wafer. The n channel device is isolated by the p well diffusion, and together with the p channel device forms a complementary pair. These devices are the basic building blocks for all CMOS logic functions, and they function as voltage-controlled on-off switches.

Two n^+ diffusions in the p well form the source and drain of the n-MOS device. The p area between these terminals is called the *channel*, which will be conducting or nonconducting, depending on the gate-to-source voltage. The gate is separated by a thin layer of oxide from the channel. Two p^+ diffusions in the n wafer form the source and drain of the p-MOS device. It also has a channel that will be conducting or nonconducting. The bulk lead of the

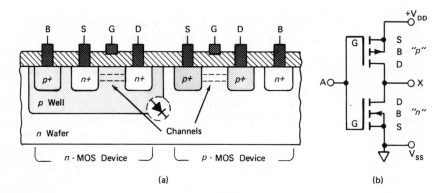

Figure 3-18. Complementary MOS Transistors. (a) Integrated structure; (b) inverter schematic.

p device is always connected to the circuits' most positive voltage, which insures that the diode formed by the *p* well and *n* wafer is reverse-biased and isolation of the two devices is established.

The complementary devices are connected as a logic inverter in the schematic of Figure 3-18(b). The upper transistor is the *p* device whose bulk lead is connected to $+V_{DD}$ and the lower transistor is the *n* device whose bulk lead is connected to ground (V_{SS}).

The input A will take on two values, $+V_{DD}$ or $0\,V$. When A is $0\,V$, V_{GS} of the *n*-MOS transistor is zero and will be off or the equivalent of an open. The *gate-to-source* voltage of the *p*-MOS device will be $-V_{DD}$. A negative V_{GS} for a *p*-MOS device will cause it to turn on hard and saturate. The output X will be a logic high of value $+V_{DD}$.

When A is high, V_{GS} of the *p*-MOS transistor is zero and will be off. The gate-to-source voltage of the *n*-MOS device will be $+V_{DD}$. A positive V_{GS} for a *p*-MOS device will cause it to turn on hard and saturate. The output X will be a logic low of value near $0\,V$.

The positive V_{GS} of the *n*-MOS transistor and negative V_{GS} of the *p*-MOS transistor must be greater, in absolute value, than the threshold voltage of each device. The threshold or turn-on voltage is the minimum value that causes the channel in the device to be induced. It typically is $2\,V$.

Virtually any logic function can be implemented by using parallel/series combinations of the *n* and *p* channel transistors. In addition, complementary MOS structures offer many advantages over circuits using the *p* and *n* channel transistors exclusively.

QUESTIONS

1. What bipolar integrated circuit components can be directly fabricated? Which is the most expensive and, hence, the least-used component?
2. How is isolation between the bipolar components in an IC achieved? What electrical condition is necessary in this technique?
3. What geometric and process characteristics define the value of a bipolar IC resistor? What technique is used to obtain large values of bipolar resistors?
4. Which bipolar transistor is most frequently used? Why?
5. List the types of *npn* transistors. List the types of *pnp* transistors. What two major functions do transistors perform in circuits?
6. What parameter describes the gain in bipolar transistors? What parameter describes the gain in MOS transistors?
7. Describe the structures of bipolar and unipolar capacitors. Which of the two is the most versatile?
8. What are the input and output variables of the *npn* transistor and the *n*-MOS transistor? Identify the directions of current and polarities of voltage.
9. What is the purpose of the *p* well in the CMOS integrated structure? How is isolation between the *p*-MOS and *n*-MOS transistor achieved? What condition is necessary to insure isolation?
10. Convert the typical vertical and horizontal dimensions in Figures 3-11(a) and 3-11(b) to mils. What multiple of a human hair's diameter is the thickness of the epitaxial layer?

4

Digital Integrated Circuits
Bipolar SSI

Arithmetic and logic operations are performed by digital circuits. This chapter discusses the least complex digital integrated circuits made with the bipolar process.

4.1. INTRODUCTION

The word *digit* refers to one of the whole or discrete numbers from zero to nine. A *digital circuit* refers to an electronic circuit that has two discrete states, described mathematically by the digits 0 and 1, and electronically as low and high (0 and +5 V typically). The voltage values between the two states carry no information and occur only during the brief transitional period going from one state to the other.

The number system that mathematically describes digital circuits is the two-digit, 0 and 1, binary system. The circuit inputs and outputs are called *variables* and equations that relate the two are called *logic equations*. The simplest electronic device used to implement a logic term is the logic gate or transistor switch. Basically its circuit is designed so that the transistor is off and the output is high, or, the transistor is on and saturated and the output is low. Both bipolar and unipolar (MOS) transistors are used for digital circuits.

A digital integrated circuit is an IC that has one or more two-state inputs and one or more two-state outputs. Usually for the basic logic functions that have one output and a minimum number of inputs, each function is duplicated as many times as the package pins will allow. For example, a 14-pin digital IC will contain six INVERTER gates, or four two-input, one-output NAND gates. Complex devices such as that used by a calculator may have many inputs and outputs and will use all the pins of a 24- or 40-pin package.

Gates and simple digital devices fall into the category of small-scale integration (SSI). *Small-scale integration* refers to a *level of complexity* of a digital IC and is gauged by the number of circuits in the IC. There are three categories, with SSI being the least complex, MSI or medium-scale integration describing moderately complex ICs, and LSI or large-scale integration encompassing the most complex circuits. Typically an SSI IC has less than 10 gate circuits, an MSI IC has from 10 to 100 gate circuits, and an LSI device has over 100 gates. The classing of ICs in this manner is only used for digital integrated circuits. Although die sizes are about the same for linear and digital ICs, linear devices have fewer transistors but have a greater number of passive components, which require a larger area on the IC chip.

Most digital ICs can be manufactured by both IC fabrication processes, bipolar and unipolar (MOS). In fact, the number of design and fabrication techniques within each process is significant; they are referred to as device families. Figure 4-1 lists a dominant member for each family and category. Our coverage of bipolar digital integrated circuits will center on those families that have the greatest impact on the digital sector of the electronics industry.

Unfortunately, there is no one logic family that possesses the best performance in all the parameters that specify a digital IC. Each family has its merits and weaknesses, and the selection of a family is a function of the

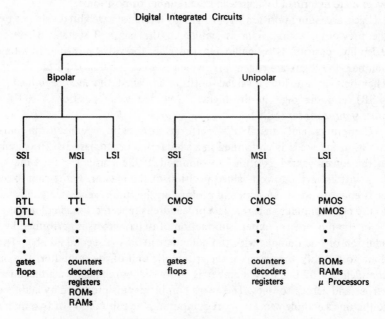

Figure 4-1. Classification of Digital Integrated Circuits

requirements of the application. These requirements include (a) flexibility, (b) speed, (c) noise immunity, (d) temperature range, (e) power dissipation, (f) noise generation, (g) drive capability or fan-out, (h) supply voltage range, and (i) cost.

DIGITAL FUNDAMENTALS

4.2. BINARY NUMBER SYSTEM

A specific amount or quantity may be represented by a symbol called a number. A different quantity may be represented by a different symbol, or a combination of symbols. Collectively, these symbols and their organization form a number system. There are many number systems with various symbols. The number of unique symbols defines the label of that system and is called its *base*.

Computers and digital integrated circuits use the binary or base 2 number system. This system has two unique symbols or digits, 0 and 1, which conveniently equate to the on and off states of a transistor in a circuit and the low and high voltages at its output. Quantities greater than 1 in the binary system are described by increasing the number of positions.

Each ascending position, from right to left, has a weighted value of twice the previous position. This is similar to the base 10 system, where each ascending position is weighted ten times the previous position. In a binary number, the digits are called *bits*, which is a contraction from *binary digits*. The digit in the lowest-valued position is called the *least significant bit* (LSB), and the digit in the highest weighted value position is called the *most significant bit* (MSB).

Computers and digital ICs perform arithmetic operations in binary. Addition in base 2 is performed in a similar manner to base 10. The numbers in the lowest-valued positions are summed. If the sum is less than the base, it is written in the lowest-valued position of the answer. If the sum exceeds or is equal to the base, a second position in the sum is generated and is taken to the next column as a carry. The procedure is repeated for all columns.

In the modern computer, subtraction of two numbers is performed by the addition of one number with the complement of the second number. This is done to simplify the design of the arithmetic unit of the computer. The one's complement of a binary number is produced by changing all the ones to zeroes and zeroes to ones. The two's complement is produced by adding one to the one's complement. In most computers, the subtraction of two numbers is performed through the addition of two numbers, one of which is in two's complement form.

The arithmetic capability of a computer begins and centers around its ability to add, and to perform a few logic operations. The computer subtracts by adding one number with the complement of the second number. It basically multiplies by successively adding, and it divides by successively subtracting. Figure 4-2 compares several key characteristics and operations of the binary and decimal systems.

	Binary	Decimal
Symbols	0 , 1	0 , 1 , 2 , 3 , 4 , 5 , 6 , 7 , 8 , 9
Positional Value	$\boxed{2^2}\,\boxed{2^1}\,\boxed{2^0}\cdot\boxed{2^{-1}}\,\boxed{2^{-2}}\,\boxed{2^{-3}}$	$\boxed{10^2}\,\boxed{10^1}\,\boxed{10^0}\cdot\boxed{10^{-1}}\,\boxed{10^{-2}}\,\boxed{10^{-3}}$
Complementation		
Number:	1 0 1 0 1	4 7 0 1
N - 1 complement	1's compl. 0 1 0 1 0	9's compl. 5 2 9 8
N complement	2's compl. 0 1 0 1 1	10's compl. 5 2 9 9

Figure 4-2. Characteristics and Operations of the Binary and Decimal Number Systems

4.3. BASIC LOGIC FUNCTIONS

Logic is the science of reasoning or argumentation. In logic, problems are reduced to simple yes and no or true and false decisions. It is customary to use the symbol 1 to represent affirmative conditions like yes, true, or present, and the symbol 0 for their opposites: no, false, and absent. Logic, as applied to digital circuits, describes the functions of these circuits and the principles under which they operate.

There are three basic digital logic operations:

1. AND: mathematically symbolized by (·) and called the *logical product.*
2. OR: mathematically symbolized by (+) and called the *logical sum.*
3. NOT or INVERT: mathematically symbolized by (⁻).

These operations and their combinations generate all others. The participants in the functions are called *variables* and are symbolically represented by letters. Since these variables have two states, they are called binary variables.

The AND function is true or exists, if and only if all the logic input variables are true. If the input variables A AND B AND C AND D are all true, then the function, or output, is true. The OR function is true or exists or is not false, if and only if at least one of the input variables is true. If the input variable A OR B OR C OR D is true, then the function or output is true.

The NOT or INVERT operation converts the state or value of an input variable to its complement at the output. If the input variable is true, then the output variable is false or not true.

Closely associated with AND, OR and NOT are the functions NAND and NOR. NAND and NOR are the NOT AND and NOT OR functions. The schematic symbols for the five basic logic functions are shown in Figure 4-3.

4.4. BOOLEAN ALGEBRA

Logic functions are mathematically described by using the *algebra of logic* called Boolean algebra. Quantitatively, the AND function, for the two input variables A and B and the output variable X, is expressed as an equation by

$$X = A \cdot B$$

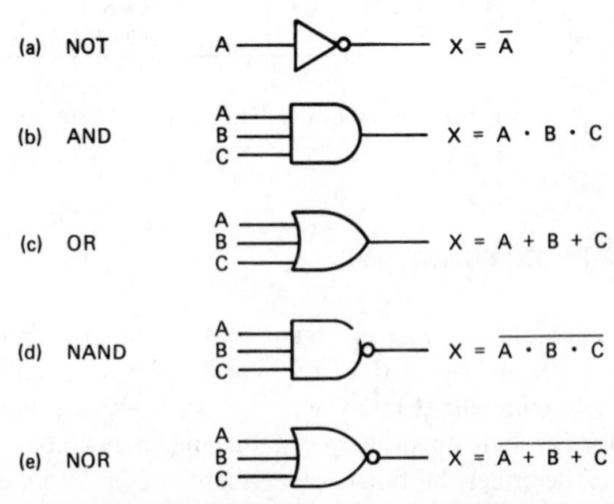

Figure 4-3. Logic Symbols for the Basic Logic Functions

If we assign numerical values, then A must be 1 and B must be 1 for X to equal 1. For any other values of A and B, X equals 0. The OR function is written as

$$X = A + B$$

A or B must be equal to 1 if X is to equal 1.

Then we have the NOT function

$$X = \overline{A}$$

If A is 1, then X equals 0. If A is 0, then X equals 1. NAND is the negation of the AND, or the NOT AND function.

$$X = \overline{A \cdot B}$$

NOR is the negation of the OR, or the NOT OR function.

$$X = \overline{A + B}$$

Figure 4-4 lists the properties, postulates, and theorems that apply to Boolean algebra.

Postulates

1. $A = 1$ (if $A \neq 0$)	$A = 0$ (if $A \neq 1$)
2. $0 \cdot 0 = 0$	$0 + 0 = 0$
3. $1 \cdot 1 = 1$	$1 + 1 = 1$
4. $1 \cdot 0 = 0$	$1 + 0 = 1$
5. $\overline{1} = 0$	$\overline{0} = 1$

Properties

6. $AB = BA$	$A + B = B + A$
7. $A \cdot (B \cdot C) = A \cdot B \cdot (C)$	$A + (B + C) = (A + B) + C$
8. $A \cdot (B + C) = A \cdot B + A \cdot C$	$A + BC = (A + B)(A + C)$

Theorems

9. $A \cdot 0 = 0$	$A + 0 = A$
10. $A \cdot 1 = A$	$A + 1 = 1$
11. $A \cdot A = A$	$A + A = A$
12. $A \cdot \overline{A} = 0$	$A + \overline{A} = 1$
13. $\overline{\overline{A}} = A$	$A = \overline{\overline{A}}$

Figure 4-4. Postulates, Properties, and Theorems of Boolean Algebra

It is possible to represent any logical function by a sum of products *or* a product of sums. This twofold or dual way of representing logic functions is the principle behind De Morgan's theorem, which states

$$\overline{A} \cdot \overline{B} \cdot \overline{C} = \overline{A + B + C}$$

and
$$\overline{A} + \overline{B} + \overline{C} = \overline{A \cdot B \cdot C}$$

Figure 4-5 lists the De Morgan equivalent for the five basic logic functions.

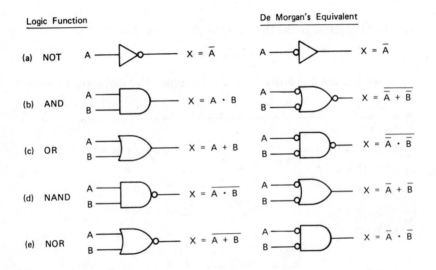

Figure 4-5. De Morgan's Equivalent of the Basic Logic Functions

4.5. TRUTH TABLES

A *truth table* diagrammatically lists the relationship of the inputs and the outputs of a digital device for all combinations. The table simply tabulates all possible conditions of input variables in a binary order. One or more columns list the outputs for each of these specific input conditions. The input and output conditions are written as 1 and 0, or high and low. Truth tables can be frequently found in the data sheets for digital ICs. The truth tables for the AND, OR, NAND, NOR, and NOT functions are given in Figure 4-6.

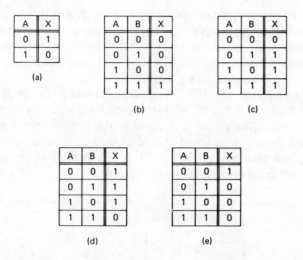

A	X
0	1
1	0

(a)

A	B	X
0	0	0
0	1	0
1	0	0
1	1	1

(b)

A	B	X
0	0	0
0	1	1
1	0	1
1	1	1

(c)

A	B	X
0	0	1
0	1	1
1	0	1
1	1	0

(d)

A	B	X
0	0	1
0	1	0
1	0	0
1	1	0

(e)

Figure 4-6. Truth Tables for the Basic Logic Functions. (a) NOT; (b) AND; (c) OR; (d) NAND; (e) NOR.

SMALL-SCALE INTEGRATION (SSI)

Bipolar device families possess the advantage of higher operating speed and greater output drive capability than their unipolar (MOS) counterparts. These attributes are particularly important to the performance of digital information processing machines or computers.

The level of complexity in a digital IC varies. The least complex category, SSI, encompasses gates and flip flops. Gates are digital circuits that implement the basic logic functions, i.e., AND, OR, NOT, NAND, and NOR. Flip-flops are digital memory cells or memory circuits.

4.6. GATES

Resistor-Transistor-Logic (RTL)

Two device families, resistor-transistor-logic (RTL) and diode-transistor-logic (DTL), have been technologically displaced, but they have played an important role in the evolutionary development of today's devices.

Prior to the development of ICs as we know them, RTL was the most popular form of logic in use. It was simple, inexpensive, and used only discrete resistors and transistors. The initial RTL ICs were direct translations from the discrete component designs, but they signaled the advent of miniaturization. The performance of the integrated circuit was comparable to the discrete circuit. Its speed was comparatively slow. It was noise-sensitive and could drive a minimum number of similar RTL gates. The gain in the shift to ICs was primarily a matter of space and economics.

The simplicity of the RTL design is illustrated in the two-input NOR circuit of Figure 4-7. The NOR circuit is the basic logic gate or function of the RTL family. The circuit is designed so that each transistor is off (open circuit) and the output is near V_{cc}, or the transistor is on and saturated (near $0V$). The two transistors are in parallel. A logic high on either one input or the other will make the output low or a logic 0.

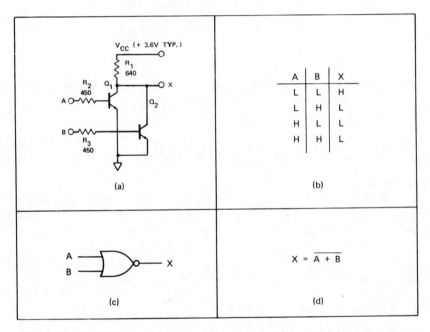

Figure 4-7. RTL NOR Gate. (a) Circuit; (b) truth table; (c) symbol; (d) logic equation.

Diode-Transistor-Logic (DTL)

The drawbacks of the RTL family were partially overcome by the introduction of the DTL or *diode-transistor-logic* family. The use of diodes instead of resistors resulted in better noise margin, fan-out, fan-in, speed, and economy. *Fan out* and *fan in* refer to the number of similar devices that can be connected to a device output and input respectively.

The NAND circuit is the basic logic gate or function of the DTL family. It consists of a diode AND circuit followed by a transistor inverter (see Figure 4-8). Logical ANDing is performed by the input diodes, D_1 and D_2. These diodes will be reverse-biased if both inputs are at 1 or HIGH with a positive signal voltage equal to V_{cc}. With D_1 and D_2 reverse-biased, Q_1 will turn on through the current supplied through R_4 and the collector feedback resistor R_1. The current of Q_1 will turn on Q_2 through D_3. Q_2 will go into saturation with the output or collector voltage near 0V or a logic 0 or LOW. R_3 is provided to remove the excess stored charge in Q_2.

De Morgan's equivalent function occurs when A is low OR B is low. If any one input drops to near ground potential, the corresponding input diode conducts and current flows through R_1 and that input diode. The voltage at the base of Q_2 is near 0.6 V, turning Q_1 and Q_2 off. The voltage at the collector rises to V_{cc} or a HIGH.

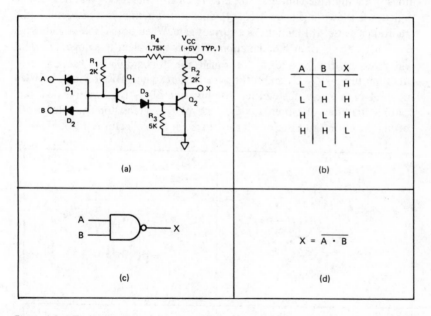

Figure 4-8. DTL NAND Gate. (a) Circuit; (b) truth table; (c) symbol; (d) logic equation.

Transistor-Transistor-Logic (TTL)

The most widely used bipolar device family is TTL, or *transistor-transistor-logic*. Its popularity is demonstrated by the many devices offered by manufacturers in the SSI and MSI categories. Only recently has bipolar technology developed to the point where LSI devices can be made. The family is charac-terized by excellent drive capability, high speed, interface compatability with other logic families, and low price. Its ability to easily generate noise and its high power consumption are major disadvantages. The TTL logic family has a wide range of applications. It is used in all segments of industry except in applications where the *ultimate* in speed, noise immunity, or low power consumption is required.

TTL NAND Gates. Figure 4-9 shows a simplified version of the TTL NAND gate. The NAND circuit is the basic logic gate or function of the TTL family. This family was created by IC technology, and the multiple emitters are its trademark or characteristic.

The simplified TTL gate has two transistors. Q_2 is a grounded emitter transistor with a collector load resistor. It is either on and saturated or com-pletely off. The on/off state of Q_2 is determined by Q_1 and the input condi-tions of its multiple emitters. The integrated structure of Q_1 and its multiple emitters is shown in Figure 4-10. Q_1 does not function as a transistor in the traditional sense. It functions as a current gate. When both signal A and signal B are high, the emitter base junction of Q_1 is equivalent to an open. Current will flow to the base of Q_2 through the forward-biased base-collector junction, turning Q_2 on and setting the output to near 0V or a logic 0. When either A or B is low, the emitter-base junction of Q_1 is forward-biased, pre-venting the flow of current to the base of Q_2. Transistor Q_2 will be off, setting the output to $+V_{cc}$ or a logic 1 through the collector resistor.

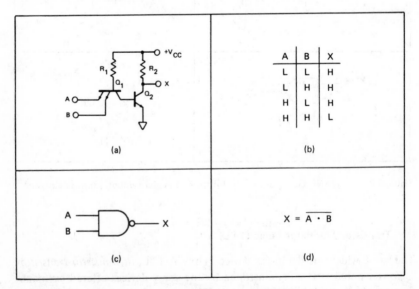

Figure 4-9. Simplified TTL NAND Gate. (a) Circuit; (b) truth table; (c) symbol; (d) logic equation.

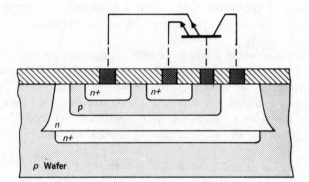

Figure 4-10. Integrated Structure of a Multiple Emitter Transistor

A standard TTL NAND gate is shown in Figure 4-11. An output stage is added to the basic circuit to increase its speed and drive capability. The second stage is also modified to drive the two-transistor output stage. These output transistors, Q_3 and Q_4, make up what is referred to as a *totem pole*. The output is high when Q_3 is on and Q_4 is off. The output going high through Q_3 is referred to as an active pull-up as opposed to the passive pull-up of a collector resistor. The output is low when Q_4 is on and Q_3 is off. For both cases, the output impedance is low, and the output stage can drive reasonably high passive and reactive (capacitive or inductive) loads. The output transistors are turned on and off by the transistor stage of Q_2, which in turn is controlled by Q_1 and the inputs A and B.

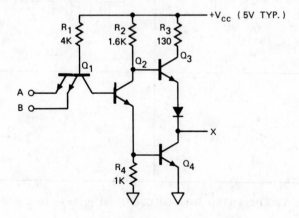

Figure 4-11. A Standard TTL NAND Gate.

The bases of the totem pole output transistors are driven by the Q_2 circuit. This circuit ensures that one of the output transistors will be on and the other will be off.

When the gate inputs A and B are high, Q_2 is turned on hard. The voltage across the 1-KΩ emitter resistor will rise, turning on Q_4 hard. As the collector current through Q_2 increases, the collector voltage drops, turning Q_3 off.

When the gate inputs A or B are low, Q_2 is turned off. Q_2's emitter voltage is near 0V, because the emitter resistor is connected to ground. This turns off Q_4. With Q_2 turned off, Q_3 is turned on through R_2, which is connected to $+V_{CC}$.

Diodes are connected to ground from each input to protect the IC from negative voltages or transients and help increase the noise margin.

The standard TTL circuit can have any number of Q_1 emitters. If the circuit has only one, it is a logic INVERTER, as shown in Figure 4-12. To maximize the usage of the IC package, the die within the package contains six inverter gates or circuits. If the TTL circuit has more than one input, the number of gates per package is reduced. Figure 4-13 shows an eight-input TTL NAND gate.

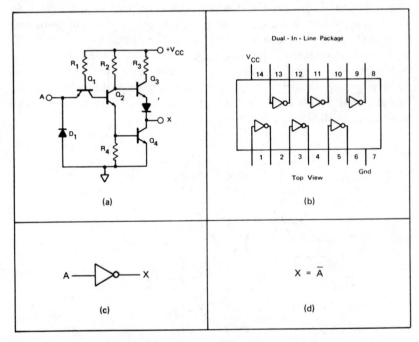

Figure 4-12. TTL INVERTER Gate. (a) Circuit; (b) package; (c) symbol; (d) logic equation.

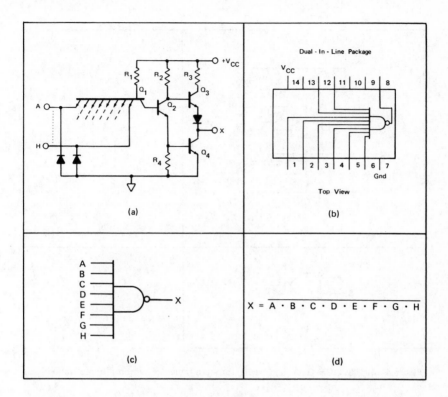

Figure 4-13. TTL Eight-Input NAND Gate. (a) Circuit; (b) package; (c) symbol; (d) logic equation.

The TTL NOR Gate. A second basic logic gate in the TTL family is the NOR gate, shown in Figure 4-14.

In the NAND gate (refer to Figure 4-11), the "ANDing" is performed with the multiple emitters of Q_1. The inversion of the signal to perform the NOT portion of the NOT AND or NAND function is established by the subsequent stages. In the NOR gate, the "ORing" is accomplished by paralleling the transistors Q_2 and Q_6. If either Q_2 OR Q_6 is turned on, the gate's output will go low. The two parallel devices have common emitters and collectors; their integrated structure is shown in Figure 4-15. Physically, the two transistors have a common collector, but their individual emitters are interconnected with metallization. A positive voltage on either the base of Q_2 or Q_6 will cause the voltage across the R_4 emitter resistor to rise and turn Q_4 on. The output will then be near 0V or a logic low. The NOT portion of the NOT OR or NOR function is established by the output stage. The input circuits of Q_1 and Q_5 function in the same way as their NAND gate counterparts.

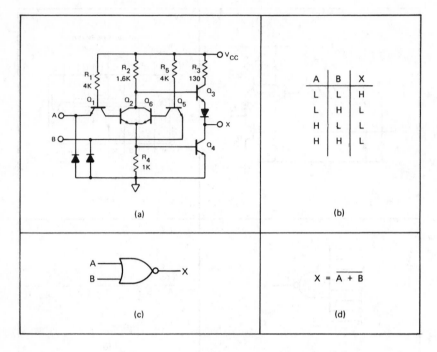

Figure 4-14. TTL NOR Gate. (a) Circuit; (b) truth table; (c) symbol; (d) logic equation.

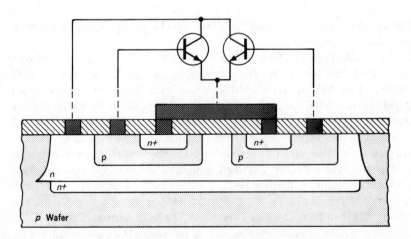

Figure 4-15. Integrated Structure of the NOR Gate Parallel Transistors

The TTL AND-OR-INVERT Gate. A simple addition to the NOR circuit will convert it to a gate that can perform the AND and OR functions. If Q_1 and Q_5 have multiple emitters, the signals on the emitters of that stage are ANDed and each type of Q_1, Q_5 stage will be ORed. The output of the OR circuit is then inverted. Figure 4-16 shows a four-input gate that is appropriately called an AND-OR-INVERT gate.

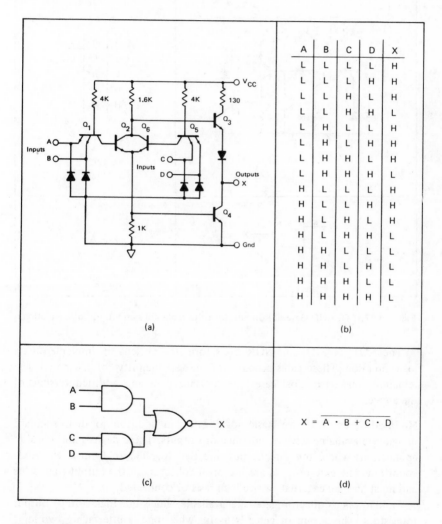

A	B	C	D	X
L	L	L	L	H
L	L	L	H	H
L	L	H	L	H
L	L	H	H	L
L	H	L	L	H
L	H	L	H	H
L	H	H	L	H
L	H	H	H	L
H	L	L	L	H
H	L	L	H	H
H	L	H	L	H
H	L	H	H	L
H	H	L	L	L
H	H	L	H	L
H	H	H	L	L
H	H	H	H	L

(a)

(b)

(c)

$$X = \overline{A \cdot B + C \cdot D}$$

(d)

Figure 4-16. TTL AND-OR-INVERT Gate. (a) Circuit; (b) truth table; (c) symbol; (d) logic equation.

The TTL AND and OR Gates. The logic AND and OR functions can be realized by adding an inverter at the output of the NAND and NOR gates. In integrated circuits, this inversion is implemented by adding an inverting stage within the IC itself. Figure 4-17 shows the AND device. This device has four transistor stages, as compared to the three stages of the NAND gate.

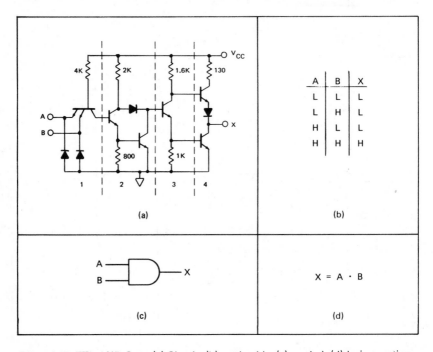

Figure 4-17. TTL AND Gate. (a) Circuit; (b) truth table; (c) symbol; (d) logic equation.

The NOT, NAND, and NOR gates form the nucleus for implementing a logic function. These gates account for the vast majority of gates used. The remaining ones, for most cases, are specialized or are modified versions of the above.

Modified Gates. The five basic logic gates, in addition to their standard design, are available with open collector outputs. These devices are useful in applications where the gate outputs are tied together to form a wire-AND condition. The gate outputs are the open collectors of the output transistors and must have an external passive load resistor connected.

Similar open-collector gates are available which use high-voltage output transistors. This group of gates is useful when one is interfacing two logic families or when one is driving relays and lamps as loads.

For many cases, the required load will exceed the output drive capability of the standard gate. To account for these applications, gates with a higher output drive capability are available; they are called buffers.

Gate Applications. Most gates are not used alone but are used in conjunction with each other to form higher-level functions. These functions may be unique to an application, or they may be commonly used logic functions. If they are used in a unique application, the NAND, NOR, and INVERTER devices are purchased as IC gates and are interconnected with discrete wires or on a printed circuit board. If the gates are used to implement a commonly used higher-level logic function, they are integrated or made as one IC and sold as that higher-level logic function. Examples include flip-flops, counters, decoders, registers, adders, converters, etc. These higher-level logic functions serve as excellent examples for the applications of gates and the building of complexity in digital ICs.

TTL Relatives

Low-Power Schottky-TTL (LS-TTL). The performance, popularity, and widespread usage of TTL has spawned four subspecies:

1. Low power: L-TTL
2. High speed: H-TTL
3. Schottky: S-TTL
4. Low-power Schottky: LS-TTL

These subspecies are designed to increase the performance of the standard TTL family in specific areas. The low-power and high-speed lines trade lower power for lower speed or higher speed for higher power. The Schottky line increases the operating speed of the devices without a significant increase in power. It derives its name from the use of the Schottky diode. This diode, connected from the collector to base of a transistor, prevents the transistor from operating in its saturation region and thus increases its operating speed.

The most recent line, technologically the best performer, is low-power Schottky, which offers a greater operating speed *and* lower power consumption than standard TTL. The same functions are offered by all members of the TTL family and are pin-for-pin compatible.

A Schottky barrier diode in parallel with the base-to-collector junction of *saturating npn* transistors forms the basis of S-TTL. The diode voltage-clamps the transistor to a value greater than V_{CE} (SAT) and diverts most of the excess base current. The transistor avoids the usual saturation, and stored charge is essentially eliminated in both the diode and the transistor. The lower forward voltage drop of the Schottky diode, compared to a *p-n* diode, provides a greater protection for the circuit under transient operating conditions.

In the integrated structure (Figure 4-18), the transistor base contact extends beyond the base (p) diffusion and over the collector (n) region. There it forms a *metal-to-silicon* or Schottky barrier diode structure. The metallization serves both as the contact to the base and as the anode of the diode. The uniqueness of the diode is indicated by the turned cathode in its symbol.

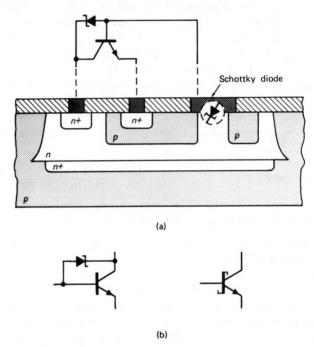

(a)

(b)

Figure 4-18. A Schottky Diode Clamped npn Transistor. (a) IC structure; (b) schematic symbols.

A complete S-TTL NAND gate is shown in Figure 4-19. The circuit is basically the same as a standard TTL gate except for the Schottky clamped transistors and the addition of Q_5, Q_6, R_4, R_5, and R_6. These devices help to give the S-TTL gate a symmetric transfer characteristic. Transistor Q_5 converts Q_3 to a Darlington pair for fast pullup. Resistor R_6 is, of course, a base bleed resistor. Transistor Q_6 and resistors R_4 and R_5 are an active pulldown circuit and replace the single resistor in the standard TTL gate.

Three-State Logic (TSL). A digital integrated circuit is defined as an IC that has one or more two-state inputs and one or more two-state outputs. The statement must be somewhat qualified to account for the TSL logic family. TSL, or three-state logic, is a modified TTL logic family whose device outputs have the standard high and low states, and also have a pseudo-third state. The

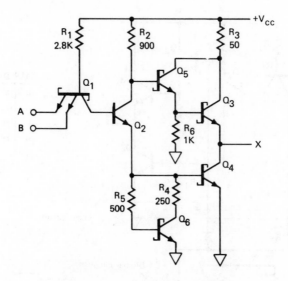

Figure 4-19. S-TTL NAND Gate Circuit

third state is not a voltage level that carries information but is a high-output impedance state that has a first-order model of an open circuit. The family was developed to overcome the problems associated with wire-AND circuits. In these types of circuits, several device outputs are connected to a digital bus line, which results in waveform integrity, speed, and loading problems. With TSL, it is possible to connect the outputs of 100 to 200 devices to a common bus line or tie point. Because of the off or third state, large numbers of three-state TTL circuits can communicate reliably with one another at high data rates, using common bus lines. Only the outputs of those devices that are doing the talking are activated. Most of the standard TTL devices are available in a TSL version. TSL is totally compatible with the TTL and DTL logic families.

The logic portion of the TSL gate is very similar to the TTL gate. $Q_1, Q_2,$ Q_4, and Q_5 (Figure 4-20) perform the same function as their TTL counterparts, except that Q_4 is converted to a Darlington pair with Q_3. The disable portion of the TSL gate is implemented with the $Q_6, Q_7,$ and Q_8 circuits. Their performance is analogous to $Q_1, Q_2,$ and an open collector Q_5. When the disable input D is low, Q_7 and Q_8 are off. The collector of Q_8 is equivalent to an open circuit, and the logic circuitry functions normally. When the disable input is high, Q_8 turns on and pulls one emitter of Q_1 and the base of Q_3 low, turning both transistors off. Q_1 turns Q_2 and Q_5 off. Q_3 turns Q_4 off, and the output sees two off transistors and a high impedance state. The truth table and logic symbol summarize the circuit's functions.

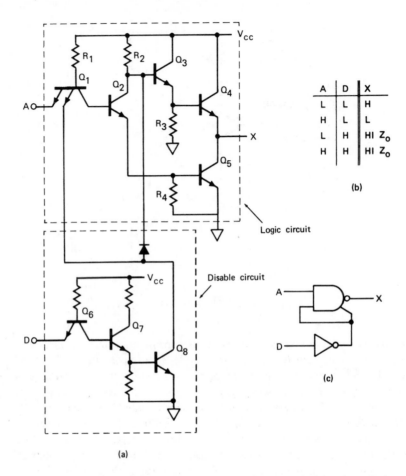

A	D	X
L	L	H
H	L	L
L	H	HI Z_O
H	H	HI Z_O

(b)

(c)

(a)

Figure 4-20. TSL INVERTER Gate. (a) Circuit; (b) truth table; (c) logic symbol.

Emitter Coupled Logic (ECL). TTL is king in the digital SSI and MSI areas. However, mention must be made of one other major IC logic family. ECL, or emitter-coupled logic, is used in applications that require the ultimate in speed. Speed is of paramount importance in high-speed digital communications systems, central processors, peripheral controllers, minicomputers, and instrumentation systems.

RTL, DTL, and TTL (except Schottky TTL) are saturated-mode logic families. They are designed so that an ON transistor will go into saturation and accumulate excess stored charge or carriers. When the transistor begins to turn off, a finite amount of time is required to remove this excess charge. In terms of device performance, the device speed is limited or reduced.

ECL is current-mode logic; the transistors do not go into saturation. While switching from one level to the other, the transistors are never totally on or saturated. The basic circuit (Figure 4-21) of the ECL 10,000 series is the two-input OR/NOR gate. Two outputs provide the complementary functions simultaneously. The basic gate consists of a differential input amplifier, a bias network, and emitter follower outputs.

The bias network, through the D_1, D_2, R_7, and R_8 divider, establishes a reference voltage (V_{BB}). This reference voltage is midway between the boundaries of the logic swing of the device. For a V_{cc} of –5 V, the outputs will swing from –2 to –4 V. When any input voltage, A or B, exceeds (is more positive than) V_{BB}, that input transistor begins conducting current. This causes the voltage at the collectors of Q_1 and Q_2 to drop and the voltage at the collector of Q_3 to rise. These voltages are directly reflected to the NOR

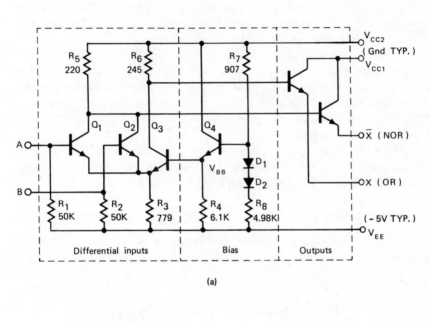

(a)

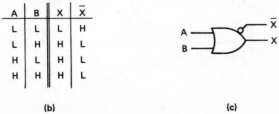

(b)

(c)

Figure 4-21. ECL NOR/OR Gate. (a) Circuit; (b) Truth Table; (c) Logic symbol.

and OR outputs through the emitter follower transistors Q_5 and Q_6. The transistors in the Q_1, Q_2, and Q_3 circuit do not go into saturation. When both inputs A and B are below V_{BB}, the collector currents of Q_1 and Q_2 cease and the level of the NOR output goes high. The low input voltages cause Q_3 to conduct, with a resulting voltage drop across R_6, and a corresponding level change occurs at the OR output. Emitter resistor R_3 serves as a current path for the switch. Transistor Q_4, diodes D_1 and D_2, and resistors R_4, R_7, and R_8 provide the temperature compensation for the reference voltage. Each input has a 50-kΩ pulldown resistor to hold unused inputs to the low state.

4.7. FLIP-FLOPS

The simplest function implemented with gates is the flip-flop, or digital memory cell. It is a device or circuit that stores or memorizes one binary bit of information. The output of the circuit has the capability to stay indefinitely in one state, a high or low, until it is told to change its state. The output state is a function of its input. Digital information may be clocked in and out of the flip-flop with a pulsed voltage, or the flip-flop can be statically operated. Most flip-flops can be set to predetermined states independent of the input conditions by using preset and clear functions. This is an important digital system requirement.

There are many types of flip-flops, but the R-S (Reset-Set), RST, D (Data), and the JK are the most important, and they serve as excellent examples to illustrate the building of complexity by using gates.

R-S Flip-Flop

The R-S flip-flop is not sold as such. It is implemented with two NAND gates. These gates are externally cross-connected, as shown in Figure 4-22.

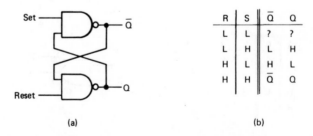

R	S	Q̄	Q
L	L	?	?
L	H	L	H
H	L	H	L
H	H	Q̄	Q

(a)   (b)

Figure 4-22. R-S Flip-Flop. (a) Circuit or logic diagram; (b) truth table.

Usually flip-flops have two outputs, labeled Q and $\overline{Q}$. The $\overline{Q}$ output is the INVERTED or NOT output of Q. The labeling or identification of the inputs is a function of the particular type of flip-flop. The inputs for the R-S flip-flop are called RESET and SET.

A low on the SET input, with RESET high, will cause the $\overline{Q}$ output to go high. If the SET input is low, the output of its NAND gate must go high. This logic high is connected to one input of the other NAND gate and along with the RESET high forces the Q output to a low. This low (Q) reinforces the high state of $\overline{Q}$. The SET input can now go high, but the flip-flop will retain its output state. This retention of output state (Q) is indicated in the truth table by repeating Q.

A low on RESET, with SET high, will cause Q to go high. For each case, $\overline{Q}$ will go to the opposite or inverted state. This flip-flop is not allowed to have two lows applied simultaneously to the inputs, since the outputs will go to an undetermined state. This is a drawback in many digital applications. The logical relationship of the inputs and outputs of flip-flops is normally shown in a diagram form called the truth table and is illustrated in Figure 4-22(b) for the R-S.

In addition to the inputs and outputs, most flip-flops have pins for auxiliary functions called CLOCK, PRESET, and CLEAR. In large digital systems, operations are performed at specific times determined by the system oscillator or CLOCK. The system clock, in conjunction with other digital signals, tells a device *when* it must perform. The system clock also synchronizes all the events that are happening either sequentially or concurrently. A "clocked" flip-flop is referred to as *synchronous* because its function occurs in relation to some fixed rate or time. A nonclocked flip-flop is called *asynchronous*; an example is the R-S type. The PRESET pin is used to set the flip-flop to a predetermined state, and the CLEAR pin is used to set the device to a start or neutral state. External logic signals control the PRESET and CLEAR pins.

RST Flip-Flop

The asynchronous R-S flip-flop can be "time" controlled by logically ANDing the R-S inputs with a signal or timing pulse T. This new flip-flop, the RST, is shown implemented with NAND gates in Figure 4-23. The TTL gates of the cross-coupled R-S flip-flop are shown using their De Morgan's equivalent NOR symbols. If the T input is low, the output state of the flip-flop remains the same independent of the R-S input states. If the T input is high, the output state of the flip-flop is determined by the R-S inputs in the same manner as before, except for the level inversion.

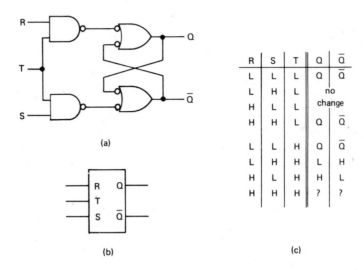

R	S	T	Q	Q̄
L	L	L	Q	Q̄
L	H	L	no	
H	L	L	change	
H	H	L	Q	Q̄
L	L	H	Q	Q̄
L	H	H	L	H
H	L	H	H	L
H	H	H	?	?

Figure 4-23. R-S-T Flip-Flop. (a) Logic diagram; (b) logic symbol; (c) truth table.

D Flip-Flop

The D flip-flop is an example of a clocked or synchronous flip-flop. As can be seen from Figure 4-24, the understanding of the circuit becomes more difficult as the complexity increases. The understanding of complex devices is greatly aided by showing the device in its block diagram form and using logic gate symbols. The D flip-flop is comprised of 3 three-input cross-coupled R-S flip flops or the equivalent of 6 three-input NAND gates. The NAND gates are simpler versions than the previously described gates, primarily because the gates within the IC do not need output buffer stages.

The D flip-flop is a digital memory cell whose output will follow its input when the pulsed clock voltage rises to a logical one. The time factor of the flip-flops operation is reflected in the truth table. A logic 0 applied to the D input at time t_n will cause the Q output of the flip-flop to go to a logic 0 at the next clock pulse, $t_n + 1$. The definitions of the logic states, i.e., 0 or 1, on which the above functions will activate may vary from manufacturer to manufacturer. Each IC device contains two or more independent D flip-flops.

J-K Flip-Flop

The J-K is the most versatile of the flip-flops. It has two data inputs, called J and K, and the standard Q and $\bar{Q}$ outputs. It is a clocked flip-flop with the preset and clear functions.

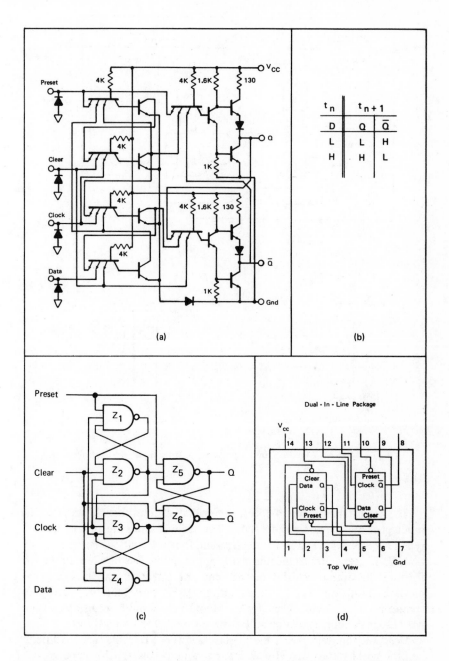

Figure 4-24. D Flip-Flop. (a) Circuit; (b) truth table; (c) logic diagram; (d) package.

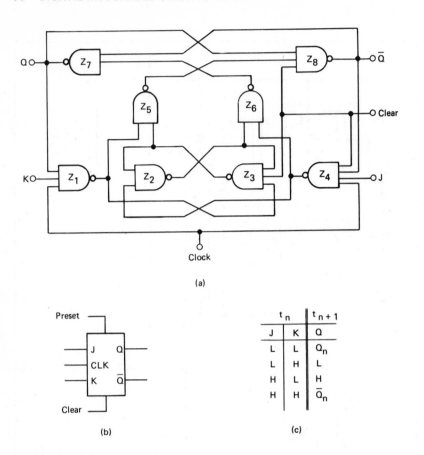

Figure 4-25. J-K Flip-Flop. (a) Logic diagram; (b) symbol; (c) truth table.

The truth table, Figure 4-25(c), shows that all four possible states of the input generate defined output states. The table is similar to that of the R-S type ($J = S, K = R$), except for the first and fourth condition. The first condition, (L–L), generates an indeterminate output in the R-S flip-flop. In the J-K flip-flop, its output remains in the same state that it was in during the previous clock time. The fourth condition, (H–H), causes the J-K to toggle or complement its output. When $J = K = H$ and a clock pulse occurs, the output will change or complement its output state from a 0 to L or a H to L.

Most J-K flip-flops are of the master-slave type. To avoid race conditions, i.e., the problems associated with internal time delays, the J-K contains two internal flip-flops. The digital information on the J-K inputs is set in the first or master flip-flop during the first half of the clock pulse. It is then strobed to the second or slave flip-flop, and hence the output during the second half of the clock pulse. In actuality, data are transferred by using the leading or falling edge of the clock pulse. Any changes of the J-K input levels do not affect the outputs during the latter part of the clock cycle.

The NAND gate representation of a J-K master-slave flip-flop is shown in the logic diagram of Figure 4-25 (a). It contains the equivalent of eight NAND gates. Devices Z_7 and Z_8 implement the slave flip-flop, and Z_2 and Z_3 implement the master flip-flop. Various circuit techniques are used in the design of the J-K flip-flop.

QUESTIONS

1. What determines the level of complexity in a digital IC? Why are linear ICs not classified in this manner?
2. What two groups of devices are classified in the SSI category? Define the function of each group.
3. What is the basic logic gate of each of the RTL, DTL, TTL, TSL, S-TTL, and ECL logic families?
4. Which logic family is the most widely used? Why? What transistor characteristic is associated with this family?
5. Identify the five basic logic functions. Which of these functions is the one most commonly found in IC form?
6. Explain the necessity of the output stage of a TTL logic NAND gate. What is this stage called and how does it function?
7. Which TTL-related family is technologically the best performer? What is its characteristic?
8. Which logic family is most appropriate to use for the parallel transmission of digital data? Which logic family is most appropriate to use in extremely high-speed arithmetic computations?
9. Which flip-flop is the most versatile? What technique is used in this flip-flop to reduce problems associated with race conditions?
10. What is meant by synchronous flip-flop? Which flip-flops are included in this category?

5

Digital Integrated Circuits
Bipolar MSI and LSI

MEDIUM-SCALE INTEGRATION (MSI)

Basically, an MSI digital IC is a device that performs a high-level logic function and has a minimum of 10 gates and a maximum of 100 gates. Most MSI devices are a combination of the basic gates previously described. This chapter discusses MSI digital integrated circuits made with the bipolar (TTL) process.

5.1. REGISTER

A flip-flop is a device or circuit that stores one binary bit of information. When flip-flops are organized to store multibit information, they are called *registers*. Registers are classified according to the way information or data are entered and removed. If all the flip-flops are set simultaneously, the register is referred to as a *parallel register*. If data are entered and removed one bit at a time, the register is referred to as a *serial* or *shift register*.

A shift register is a group of cascaded flip-flops. Each flip-flop output is connected to the input of the following flip-flop, and a common clock pulse is applied to all flip-flops, clocking them synchronously.

The MSI IC shown in Figure 5-1 is an 8-bit serial-in, parallel-out shift register. The device has one input and eight outputs. Its input will accept, at a clocked rate, eight digital 1's and/or 0's in serial form. The eight outputs will be set to a logic 1 or 0, depending on the input state at that clock time. As an example, consider an eight-bit stream of digital information clocked into the device with eight clock pulses. If the input was at a logic 1 at clock time 1, Q_1 will be set to a logic 1 or a high. The device is called a shift register

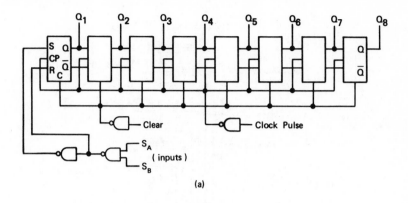

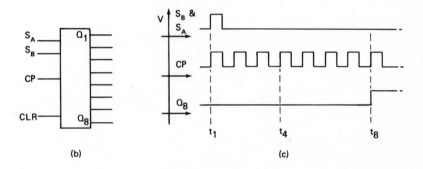

Figure 5-1. Shift Register; Eight-Bit Serial-In, Parallel Out. (a) Logic diagram; (b) logic symbol; (c) a shifted bit 1.

because the logic 1 or 0 that occurred at clock time 1 is shifted to output 8 (Q_8) after it has been clocked through the other seven flip-flops. The circuit is designed to bounce the 1 or 0 to the next series flip-flop when the next clock pulse occurs. If the clock continually runs, the register will hold the last eight bits in a digital stream of information. The flip-flops are similar to the *RST* flip-flops previously described.

A logic high on the inputs S_A and S_B will be entered into the register on the leading edge of the clock pulse *CP*. Subsequent clock pulses will advance the memorized logic one. The clear signal must be low for the above conditions to occur. If, at any time, the clear input is brought high, the register or Q outputs will go to a logical 0 or low.

Similar registers are available that can shift the stored information to the left or the right. Other versions provide for serial in–serial out, parallel in–serial out, or parallel in–parallel out operating modes.

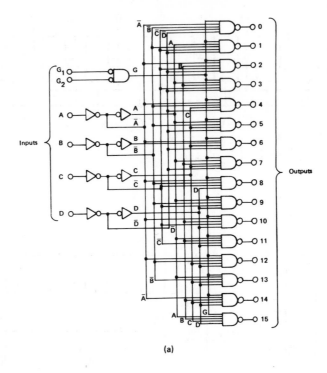

(a)

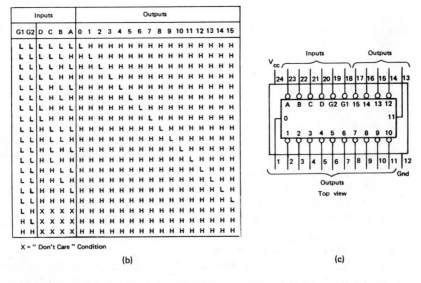

X = " Don't Care " Condition

(b)

(c)

Inputs
Outputs
Top view

Figure 5-2. Four-Line to 16-Line Decoder/Demultiplexer. (a) Logic diagram; (b) truth table; (c) package.

5.2. DECODER

A decoder is a device that *converts* or decodes digital information from one format to another. The MSI decoder shown in Figure 5-2 converts the four-bit binary coded inputs into one of 16 separate outputs. The device has two STROBE lines, both of which have to be in the low state in order to perform the decoding or converting function. The strobe lines, G_1 and G_2 behave similarly to an ENABLE function. The four binary inputs together have 16 possible combinations of 1's and 0's. Each one of these particular combinations makes *one* unique output go low. The device is the equivalent of 16 NAND gates, 8 INVERTERS, and a NOR gate.

The decoding of the 5 output for a 0101 input is shown in Figure 5-3. The G_1 and G_2 inputs are strobe or enable control lines. They both must be in the low state for the decoding function to occur. If either of these signals is high, all decoder outputs will go high independent of the input conditions. Gate Z_1 is a five-input NAND gate whose inputs must be high for the output to go low. If G_1 and G_2 and B and D are low, and A and C are high, the output will be low. All inputs are buffered with inverters to prevent loading problems for the NAND gate. The device can also be used as a demultiplexer by passing information from one of the strobes (the other being low) to an output selected by the four-input address.

The 4-to-16 decoder is just one of many types. Other noteworthy types include the 16-to-1, 8-to-1, and 4-to-1 multiplexers, the 1-to-8 and 2-to-4 demultiplexers, and the *BCD* to decimal and seven-segment decoders. All of these devices use the basic gates to transform the digital information on their inputs to digital information of a different format at their outputs.

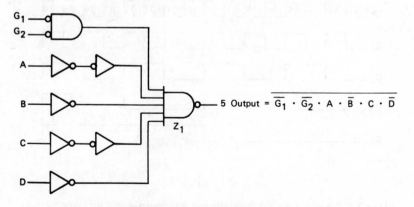

Figure 5-3. Decoding a 5 in the Decoder/Demultiplexer IC

5.3. COUNTER

A counter is a device that counts and remembers the number of clock pulses that have been applied to its input. The memory devices within the counter are flip-flops.

The counter in Figure 5-4 is a basic counter implemented with, or in, integrated circuits. It is a simple 4-bit binary ripple counter and uses J-K flip-flops for counting and memory. The Q output of each J-K is connected to the clock input of the next J-K. The J-K inputs for all flip-flops are tied to a logical 1.

Initially the Q outputs of all flip-flops are set to a logical 0 state. A clock pulse is applied to the clock input of the first flip-flop and causes Q_1 to change from 0 to 1. Since the flip-flops in the ripple counter are not under the command of a single clock pulse, it is an asynchronous counter. Flip-flop 2 does not change state, since it is triggered on the negative-going edge of the clock pulse. With the arrival of the second clock pulse to flip-flop 1, Q_1 goes

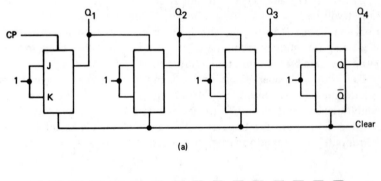

(a)

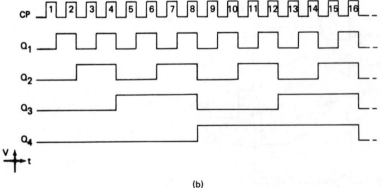

(b)

Figure 5-4. Counter, Four-Bit Binary Ripple. (a) Logic diagram; (b) counting clock pulses in binary.

from 1 to 0 and this triggers flip-flop 2 with its output Q_2 going from a 0 to 1. The four Q outputs count, in binary, the number of clock pulses. Clock pulse 16 causes the four flip-flops to go to 0 and count anew. After ten clock pulses are applied to the counter input, the outputs Q_2 and Q_4 will be high and Q_1 and Q_3 will be low.

Additional flip-flops can be added to increase the maximum count. These types of counters have $2N$ discrete states (N = number of flip-flops) and have a maximum count of $2^N - 1$. The remaining count is 0000. Versions are available that decode the outputs to a format other than binary; the most popular of which is BCD, or binary coded decimal. Counters are also available with preset, reset, count-up, and count-down features.

5.4. ADDER

The essential function of most computers is to perform arithmetic operations, and almost all computers use arithmetic elements. The fundamental element is the binary adder.

When two digits in a column are added, the resultant may contain one or two digits. If the result of the addition *does not* exceed the highest number in its number system—for binary a 1—the answer is the sum and the second digit, called a *carry*, is a zero. If the result of the addition *does* exceed the highest number, the second digit of the answer is a one and must be added to the next-highest-ordered column. The logic diagram of Figure 5-5 is for a half-adder and illustrates the sum and carry ideas. It is called a *half-adder* because it contains a carry-out but not a carry-in feature. A half-adder may be used for adding two numbers in the first column only. The logic equation for the sum S is known as the exclusive-OR function and is similar to the OR

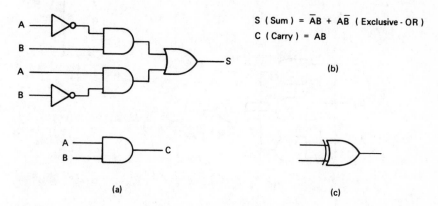

$$S \ (\text{Sum}) = \overline{A}B + A\overline{B} \ (\text{Exclusive - OR})$$
$$C \ (\text{Carry}) = AB$$

(b)

(a)

(c)

Figure 5-5. Half-Adder. (a) Logic diagram; (b) logic equations; (c) exclusive-OR symbol.

function except for the 1-1 input states. This modified OR function has its own logic symbol. The sum S will be a logical 1 for A equals 1 and B equals 0, or A equals 0 and B equals 1. The sum S will be logical 0, for A and B equal to 1, or A and B equal a 0. The carry C will be a logical 1 for A and B equal to a 1.

Figure 5-6 contains the logic diagram for a single-bit full-adder. Its inputs are the two numbers and a carry-in, and its outputs are the sum and a carry-out. The carry-in is connected to the next-lowest-ordered adder (column), and its carry-out is connected to the next-highest-ordered adder (column). Single binary bit adders are connected together to form multibit adders. Figure 5-7 shows a four-bit parallel adder. The binary number $A_3A_2A_1A_0$ is added to the binary number $B_3B_2B_1B_0$ to form a sum $S_3S_2S_1S_0$ and carry C_{O4}. Adder ICs are usually 4-bit full-adders. Several are usually used together to handle the computer system's data word.

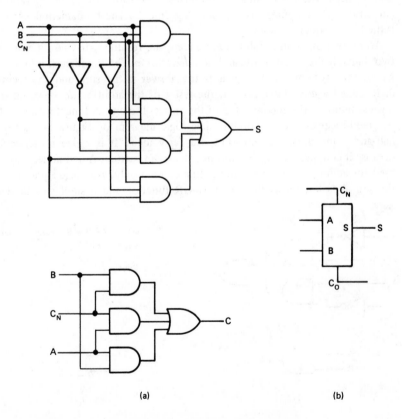

(a)   (b)

Figure 5-6. Full-Adder. (a) Logic diagram; (b) logic symbol.

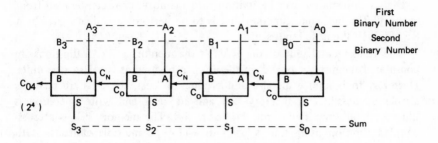

Figure 5-7. Parallel Adder, 4 Bits

5.5. RANDOM ACCESS MEMORY (RAM)

A RAM, or *R*andom *A*ccess *M*emory, is an array of memory cells where digital information is inputted, stored, and outputted using commands and addresses. A more proper name for RAM is Read/Write Memory, because data are accessed in the same manner for both functions. Digital data, i.e., organized 1's and 0's, may either be written into memory or read from memory repeatedly. RAM IC's are used collectively to form a storage medium in computers and digital processing machines. In computers, they are used for main memory, for smaller memories associated with arithmetic operations called "scratch pad memories," and for buffer memories used for input-output units. In each of these applications, data are stored in memory and later retrieved for subsequent processing. Data storage in all semiconductor read/write memories is volatile, i.e., data can be stored only as long as power is uninterrupted. Memory *subsystems*, implemented with a number of IC's, are generally identified by number of words, number of bits, and function. For example, a 1024×16 RAM is random access read/write memory containing 1024 words of 16 bits each. Semiconductor memory *device* organizations follow the same rule. A 16×4 RAM IC is a random access read/write memory containing 16 words of four bits each. Individual RAM IC's are expandable to allow each to operate in conjunction with other devices to form the complete memory subsystem. Most devices contain address decoders, output sensing, and various control and buffer/driver functions in addition to the array of storage cells. High-density RAM devices tend to be organized in *n* words by one bit to optimize lead usage.

Each memory cell must have a unique address. The digital bits in an address locate a specific cell by identifying its row and column. To minimize the number of bits, they are usually decoded. The same addressing technique is used for reading or writing into memory.

Digital information can be written into memory, or it can be read from memory. The particular function that is to be performed is determined by a READ/WRITE enable input signal.

The fundamental unit of the RAM is the memory cell. All flip-flops are considered memory cells and, in theory, can be used as RAM storage units. However, to minimize die size and production cost, the cells are made as simple as possible. A memory cell and its read and write operation is illustrated, in simplified form, in Figure 5-8. The memory cell is a cross-coupled flip-flop using transistors with TTL-type multiple emitters. Both outputs, 0 and 1, may be sensed in this type of cell. To activate the cell for a read or write operation, the $X_O Y_O$ address lines must be made a logic high. This essentially eliminates the operation of two of the emitters in each transistor. A 0 is written into memory by bringing the WRITE 0 line high. This turns on Q_5 hard and provides an active 0V ($\simeq$ 0.2V) for Q_1's third emitter. Q_1 of the cross-coupled flip-flop is driven into saturation, and Q_2 is turned off. Memory is maintained by providing a current sink of Q_1's emitter current either through the base of Q_6 or a logic low of the address lines. The grounding of Q_2's emitter does not affect its off status. To READ a 0, the WRITE line is, of course, low and Q_5 is off. The $X_O Y_O$ address lines are brought high, and Q_1's emitter current is sinked by the base of Q_6. Q_6 is turned on and its collector is brought low. In the 1-SENSE circuitry, Q_2 and Q_3 are off. Q_4 is also off because the base of this transistor is returned to ground through R_2. The READ 1 output voltage is near $+V_{CC}$ sensed through the pullup resistor.

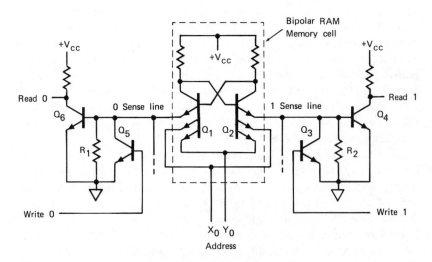

Figure 5-8. Bipolar RAM Memory Cell

The sense lines of all memory cells are connected together. One READ and WRITE circuit (or two) is used, but each cell's address lines are unique. These circuits may or may not be included in the IC. An expanded 16 × 1-bit memory block is shown in Figure 5-9. The cells are drawn in a 4 × 4 block for the 4 × 4 addressing scheme.

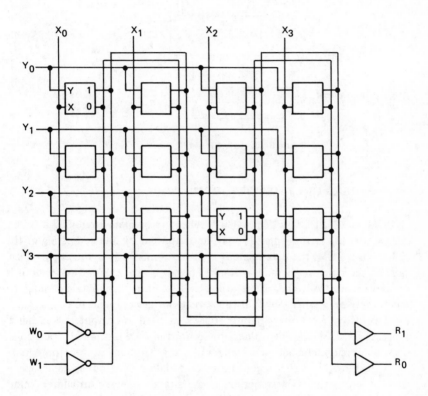

Figure 5-9. 16 x 1 RAM Memory Block

The block diagram of Figure 5-10 is for a typical RAM IC. Four data bits are either read into memory or retrieved from memory. The memory matrix has 16 rows of four bits each. The particular group of four bits that is to be processed is determined by the address bits. The address bits, four binary digits, are decoded so that only one of the decoder's output lines is activated, and hence, one group of four bits in memory is accessed.

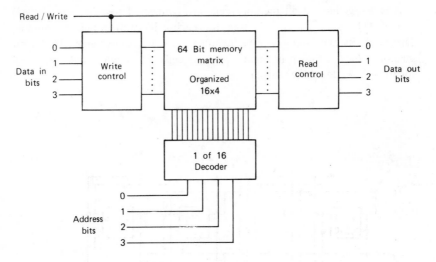

Figure 5-10. 16 x 4 RAM IC Block Diagram

5.6. READ ONLY MEMORY (ROM)

A ROM, or *R*ead *O*nly *M*emory, is a collection of preprogrammed memory cells whose digital information may be outputted by the addressing of the cell(s). In a ROM, the data content is fixed, normally by a unique metallization of the chip. Addressing is similar to that of a RAM; and its readout is nondestructive. A ROM offers nonvolatile storage; i.e., data are retained indefinitely even when power is shut off. ROM devices tend to be organized in *n* words by four or eight bits. ROM's are used in computers and digital processing machines, where their applications include arithmetic operations, code conversion, character and random logic generation, and microprogramming. In most cases their application is a variation of a lookup table operation, in which the ROM functions as a digital dictionary. An address input word becomes a reference to locate a new (and bigger) word.

A ROM cell may be programmed in one of two ways. The first method is accomplished during the fabrication process. A link, or connection, is either made or left opened during the metallization step. The presence or absence of the links is specified by the user or customer, who submits to the IC manufacturer a table defining the logic state for the memory cell at each address. The second method allows the user to program the device as he desires at his facility. Special devices, with links made of deposited Nichrome, are required. A 1 is stored in a cell by addressing the cell and pulsing it with a high current, which causes the link to be blown open. Otherwise the cell's content remains a logical 0. These devices are called field-programmable ROM's or PROM's.

The fundamental unit of the ROM is the memory cell. Bipolar ROM cells, Figure 5-11, use *npn* transistors and multiple emitters. The two cells shown, with addressing and read circuitry, represent a portion of a 4 × 4 ROM. Transistor Q_1 is the address gate, and the multiple emitters of Q_3 contain the output word bits. To increase the number of memory cells, the number of Q_1 emitters must be increased. To increase the number of bits in the output data word, the number of Q_3 emitters must be increased.

A ROM cell is addressed by bringing the XY lines high. With both X_0 and Y_0 high, Q_1 is off. Transistor Q_2 turns on through R_1, turning Q_3 on, and the emitters of Q_3 go to near $+V_{CC}$. If an emitter program link is connected, that emitter will feed $+V_{CC}$ to the R_2-R_3 resistor divider and turn Q_4 on hard. The output voltage of Q_4 is a logic 0. If an emitter program link is opened, Q_4 turns off because its base is returned to ground through R_3, and the output voltage of Q_4 is then a logic 1.

If either $X_0 Y_0$ address line is low, a base-to-emitter junction of Q_1 will be forward-biased. The current through R_1 will be sinked by the address line, turning Q_2 and Q_3 off, and deactivating the memory cell.

The block diagram of Figure 5-12 is for a ROM with a 256-bit memory matrix containing 32 words of 8 bits each. Five bits are required to address the 32 words. These address bits are decoded in a 1 of 32 decoder. Only one

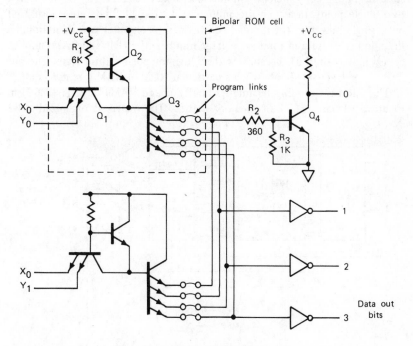

Figure 5-11. Bipolar ROM Memory Cells

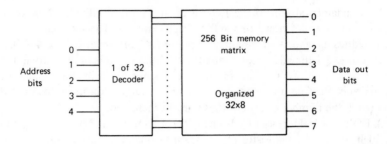

Figure 5-12. ROM IC Block Diagram

of the decoder's 32 output lines will be activated, and the specific one will be a function of the binary coded address bits. The activated decoder line will enable a specific group of eight bits to be read at the data-out lines.

Many ROMs, and RAMs, have features other than those discussed. Included among them are memory enable inputs, buffers and drivers, latched outputs, open collectors for expansion, and three-state outputs. Devices with three output states were discussed under TSL. ROMs and RAMs with three-state outputs greatly increase the number of devices that may be connected to a single point. In a typical computer system, digital data are transmitted on a single bus, with the number of bits in the computer word determining the number of lines in the bus. A large number of ROMs and RAMs may be connected to the bus, because of their high-impedance third state. The bus lines become activated only when a particular ROM or RAM is sending data.

The address decoding techniques in ROMs and RAMs vary. The ROM in Figure 5-13 has a 1024-bit memory matrix. The matrix is organized as 32

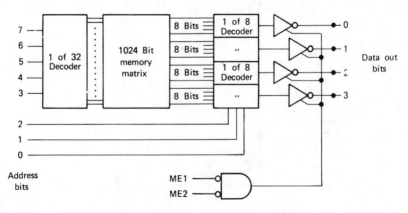

Figure 5-13. 256 x 4 Bipolar ROM Block Diagram

words of 32 bits each, but the ROM is organized as 256 words of four bits each. This is accomplished through two levels of address decoding. The five most significant address bits are used, through a decoder, to release 1 of 32 words of 32 bits each. The three least significant address bits, through four decoders, release four specific bits of the 32. These output bits are connected to the inputs of four TSL inverters, whose output states are controlled by the *M*emory *E*nable signals 1 and 2. When both enable inputs are low, the inverter outputs present the contents of the word selected by the address inputs. If either or both of the enable inputs is high, all four device outputs go to the OFF or high-impedance state. The TSL outputs of the ROM can be connected to many other ROMs with minimal loading effect. Nearly all bipolar ROMs and RAMs are TTL-compatible.

5.7. PROGRAMMABLE LOGIC ARRAY (PLA)

A PLA, or *P*rogrammable *L*ogic *A*rray, is an array of logic elements, i.e., AND, OR, and INVERTER gates, such that a given input function produces a known digital output function. The functional relationship of the inputs and outputs is specified by the user or customer, who submits to the IC manufacturer a table. This table defines the interconnections of the logic elements in the PLA. A special metallization mask is then generated to program the custom device. PLAs, typically, have 14 data inputs and 8 outputs. The AND gate outputs are called *product terms* , i.e., $A \cdot B \cdot C$. Each PLA output provides a sum of these product terms, i.e., $A \cdot B \cdot C + D \cdot E \cdot F + G \cdot H \cdot I$, where each product term can contain any combination of the 14 input variables or their complements. The total number of product terms that can be provided is 96. The PLA logically can be described as a collection of "ANDs" that may be "ORed" at any of its outputs. Figure 5-14 shows the logical data flow from the 14 input terminals through the "AND" gates to the "OR" gates and to the output. The input variables, or their complements, are combined (ANDed) to form the partial product terms. Logically any or all of the partial product terms (AND terms) can be combined (ORed) at each output. The outputs can be in their complemented or noncomplemented form.

A PLA may be programmed in one of two ways. The first method is accomplished during the fabrication process. A link, or connection, is made or left opened during the metallization step according to customer specifications. The second method allows the user to program the device at his facility. Special devices, with links made of deposited Nichrome, are used. The link can be left intact or opened by pulsing it with a high current. These devices are called Field Programmable Logic Arrays, or FPLAs.

PLAs are used to replace combinational logic circuits employing discrete IC gates. The combinational logic of Figure 5-15 is an example of a circuit

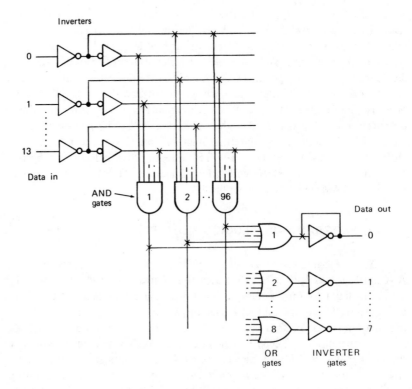

Figure 5-14. Logic Diagram for a PLA

that can be implemented with a portion of a PLA. The logic equation shows the sum (OR) of product (AND) terms. The hard wired logic must be partitioned into a relatively small number of input (14) and output (8) functions to match the capability of the PLA. The PLA is then programmed to duplicate the input-output logic relationships. It can replace as many as 25 IC gates. The PLA offers a user the option of further integrating his custom design implemented with discrete IC AND, OR, and INVERTER gates. Each time a system's physical size can be reduced by use of more complex elements such as a PLA, the cost of that system also decreases, because less hardware and labor are required. Other applications of the PLA include code converters, sequential controllers, and digital processors.

Many applications can use a PLA *or* a ROM. The ROM produces a digitally coded output for a digitally coded address input. The two are logically related. The PLA can be viewed as a memory storage device, i.e., a limited-

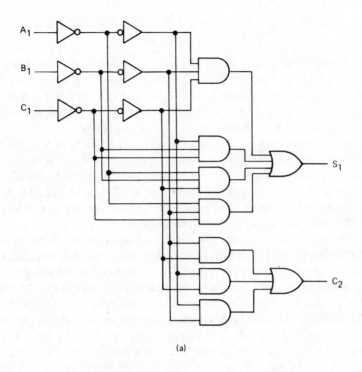

(a)

$$S_1 = A_1 \cdot B_1 \cdot C_1 + A_1 \cdot \bar{B}_1 \cdot \bar{C}_1 + \bar{A}_1 \cdot \bar{B}_1 \cdot C_1 + \bar{A}_1 \cdot B_1 \cdot \bar{C}_1$$

$$C_2 = B_1 \cdot C_1 + A_1 \cdot C_1 + A_1 \cdot B_1$$

(b)

Figure 5-15. Combinational Logic Implemented with a PLA. (a) Combinational logic diagram; (b) logic equations.

capability ROM, because its digitally coded inputs produce outputs in accordance with a predefined set of rules. Simply, both devices translate input codes of one type into output codes of another type. The number of input variables represents the key selection factor between the two. When more than nine input variables are involved, the PLA generally becomes the economical choice.

The number of devices within the MSI category constantly increases. New devices enter the market as applications call for their usage. Their functions are diverse and vary from control to memory to arithmetic operations.

L A R G E - S C A L E   I N T E G R A T I O N   (L S I)

The importance of MSI and LSI lies in the economics of its application. Although a single MSI or LSI device is more expensive than a single-gate IC, the MSI or LSI device is usually less expensive than the equivalent number of gates that are needed to implement the higher logic function. More important than the cost of the ICs is the savings in material and labor.

LSI devices in the bipolar processing class are not devoid. For the moment, unipolar processing (MOS), with its smaller transistors and die, is more practical and less costly. New bipolar technologies, notably I^2L (Integrated Injection Logic) seriously challenge the dominance of MOS in the LSI field. Both technologies, bipolar and unipolar, have their advantages and disadvantages. They ultimately will share the market in all categories.

The requirement for a high operating speed in a complex IC has pushed the development of I^2L. Clock, calculator, computer, and memory LSI circuits are already being made using I^2L techniques. The operating speed, power consumption, and interface compatability are comparable to other existing bipolar technologies but the density is one-fifth to one-tenth as great. This is primarily achieved by eliminating the component isolating *pn* junctions. In a digital device, the I^2L gate takes up the room of a single multi-emitter transistor. The gate circuit is the equivalent of a direct coupled transistor logic gate and utilizes active loads to avoid the large ohmic resistors.

QUESTIONS

1. What factor determines whether a digital IC is classed as a SSI or MSI device?

2. How many half-adders are used in a 16-bit computer arithmetic unit? How many full adders are used?

3. Which input states are similar and which are dissimilar between the OR and EXCLUSIVE-OR functions? Describe the differences of the logic symbols.

4. What are ROMs called when they can be programmed at the user's facility? Describe the technique of customer programming.

5. What are PLAs called when they can be programmed at the user's facility? Describe the technique of customer programming.

6. Does the presence or absence of an emitter program link in a ROM cell establish a logic 1 output? For this condition, trace the signal path from the output back to the initial condition that established it.

7. What are the four basic operating modes for registers? What is the operating mode of the shift register?
8. Is the binary ripple counter synchronous or asynchronous? Why?
9. What is the difference between a ROM and RAM? Which device is more appropriate to use as a code converter?
10. Describe the ROM and RAM memory cells. Identify their similarities and differences.

PROBLEMS

1. Draw that portion of the decoder in Figure 5-2 that is necessary to decode a 14. Write the logic equation.
2. What is the maximum count for a 10-bit binary ripple counter? Define the output states for a count of 646.
3. What MSI device does the PLA circuit of Figure 5-15 simulate? How many of these circuits can one PLA implement?
4. Two binary numbers, 1110 and 1001, are added in the parallel adder of Figure 5-7. Define the output state of the sum $S_3 S_2 S_1 S_0$ and the carry C_{O4}.
5. How many four-bit parallel adders are required to add the decimal numbers 1056 and 832?
6. If the memory of a computer main frame contains 10,000 16-bit words, what is the minimum number of bits required to directly address each word?
7. What is the minimum number of 16 $\times$ 4 RAM ICs necessary to implement a memory subsystem of 10,000 words of 16 bits each?
8. Define the output states of the counter in Figure 5-1(a) for S_A, S_B, and Clear set to a logic high and CP clocked four times.
9. If $X_0 = 0$ and $Y_0' = 1$ in the ROM cell of Figure 5-11, what are the nominal base voltages of Q_2, Q_3, and Q_4?
10. If $X_0 = 1$, $Y_0 = 1$, Write 1 = 1, and Write 0 = 0 in the RAM cell of Figure 5-8, what are the nominal base voltages of Q_6, Q_1, Q_2, and Q_4?

6

Digital Integrated Circuits
Unipolar (MOS)

Arithmetic and logic operations are performed by digital circuits. This chapter discusses the digital integrated circuits made with the unipolar process.

6.1. INTRODUCTION

Unipolar monolithic ICs pertain to those devices whose components are primarily *voltage controlled* and have a single polarity operational current in their active elements. They are divided into metal-oxide semiconductor (MOS) and junction field-effect transistor (JFET) categories according to the type of transistor made. Junction field-effect transistors have limited application in ICs and are mostly found as discrete transistors. The number of classes within the monolithic MOS IC category is large and steadily increasing as manufacturing technologies develop.

Figure 6-1(a) illustrates three of the major MOS classifications by device type. Analogous to the *npn* and *pnp* bipolar transistors are the *n*-MOS and *p*-MOS field-effect transistors. These devices normally operate in the enhancement mode; i.e., it takes an *increasing voltage* to cause them to conduct harder. The *n* and *p* refer to the device's type of channel or the polarity of its carriers. CMOS, or complementary MOS, refers to the class that combines the *p*-MOS and *n*-MOS devices.

Unipolar device families possess the advantage of low power consumption, and smaller transistor and resistor die size. Low power consumption is important in portable, battery-operated applications. The smaller die size of MOS is important in the economics of the manufacturing of LSI devices.

The number of device families using the unipolar process is large. However, CMOS, *n*-MOS, and *p*-MOS technologies dominate their respective categories

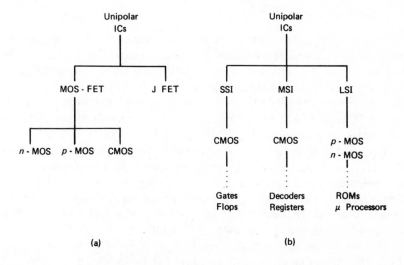

Figure 6-1. Classifications of Unipolar Monolithic ICs. (a) Type of transistor; (b) level of complexity.

[Figure 6-1(b)] and continually extend their influence. *P*-MOS and *n*-MOS are used extensively in making LSI devices and are mature technologies. CMOS is relatively new but possesses technical advantages that neither bipolar nor other unipolar families can match.

SMALL- AND MEDIUM-SCALE INTEGRATION (CMOS)

6.2. SMALL-SCALE INTEGRATION (SSI)

The most promising unipolar device family in the SSI and MSI classes is CMOS or complementary MOS. Its popularity is demonstrated by the many ICs offered by manufacturers that are functionally equivalent (and sometimes pin-for-pin the same) to the older and popular bipolar devices. The family is characterized by low power consumption, small die size, and high noise immunity. Noise immunity, a measure of the ability of a device to reject electrical interference, is extremely important in industrial control applications. The high degree of noise immunity in CMOS is not found in any other unipolar or bipolar family. The moderate difficulty of interfacing CMOS with other logic families is probably its greatest disadvantage.

CMOS Inverter

The basic CMOS circuit is the inverter shown in Figure 6-2. It consists of two MOS enhancement mode transistors, the upper a p-channel type, the lower an n-channel type. The power supplies for CMOS are called V_{DD} and V_{SS}, or V_{CC} and Ground, depending on the manufacturer. V_{DD} and V_{SS} are carry-overs from conventional MOS circuits, and V_{CC} and Ground are carry-overs from TTL logic. CMOS devices are typically powered with a single +5 V supply, but supply potentials as high as 18 volts may be used.

The input A of the inverter is connected to the two isolated gate leads of the MOS transistors. The input current to these devices is extremely small, and hence the input impedance is high, typically 10^{12} Ω in parallel with a 5-pF capacitor. Depending on the input, one of the two transistors will be on and saturated. The output impedance will be low, and the logic levels seen in a CMOS system will be essentially equal to the power supplies.

The fanout, or output drive capability, of a CMOS gate is very high. Typically, the output of one gate can drive the inputs of 50 other gates. The n- and p-channel transistors are designed with a threshold or turn-on voltage of approximately 2 volts. This causes the switching action to be sharp and to occur when the input voltage equals about half of V_{DD}. The noise immunity is also high, about 40 percent, when V_{DD} is 5 V.

The integrated structure of the CMOS inverter is shown in Figure 6-2(b). The n-channel device is isolated by the p-well and together with the p-channel device forms a complementary pair. A conductive path or channel will be induced between the source and drain of each transistor, depending on the

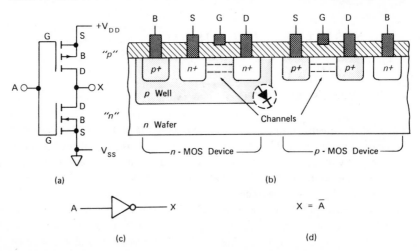

Figure 6-2. CMOS Inverter. (a) Schematic; (b) integrated structure; (c) symbol; (d) logic equation.

gate-to-source voltage. The fourth lead for each transistor is called the *bulk* or *substrate*. The bulk for the *n*-MOS device is the *p*-well and is returned to ground. The bulk for the *p*-MOS device is the *n*-wafer, and it is returned to $+V_{DD}$. These connections cause the devices to be electrically isolated. The direction of the arrow in the bulk lead of the transistor symbol indicates the type of channel; in for *n* and out for *p*.

NAND Gate

The CMOS equivalent of the TTL NAND gate is shown in Figure 6-3. Both the TTL and CMOS gates perform the same *logic* function.

Transistors Q_1 and Q_2 are *p*-channel MOS devices, and Q_3 and Q_4 are *n*-channel MOS devices. Both types of devices operate in the enhancement mode; i.e., a more positive or more negative voltage on the gate will cause the device to conduct more drain current. It should be noted that Q_1 and Q_2 are in parallel and Q_3 and Q_4 are in series. Since the device is a NAND gate, its output X will be low for both inputs A and B high. When B is high (V_{DD}),

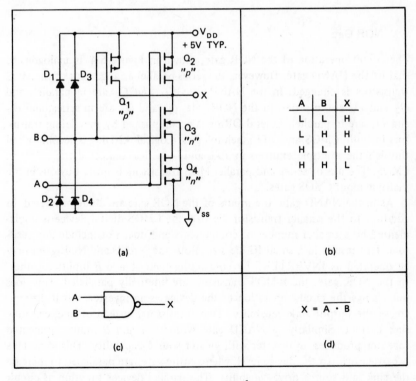

Figure 6-3. CMOS NAND Gate. (a) Schematic; (b) truth table; (c) symbol; (d) logic equation.

the gate of Q_1 is 0V with respect to its source (V_{DD}). Since the device operates in the enhancement mode, the device will be off. The gate of Q_4 is positive with respect to its source (0V) and will turn on hard. When input A is high (V_{DD}), the gate of Q_2 is 0V with respect to its source (V_{DD}). The device will be off. The gate of Q_3 is positive with respect to its source (0V through Q_4) and will turn on hard. The output X is connected to ground through Q_3 and Q_4, which are both turned on hard. Thus, X, the output, will go low (0V) when inputs A and B are high (V_{DD}). Diodes D_1 through D_4 are used to protect the inputs of the high-impedance CMOS device. They insure that the inputs will be no more positive than one diode drop above V_{DD} nor more negative than one diode drop below ground (V_{SS}).

If either A or B is low, the output will go to a logic high. For A low, the gate-*to*-source voltage of Q_2 will be highly negative and will cause the p device to turn on hard and set the output to $+V_{DD}$. Q_1 is in parallel with Q_2 and similarly will cause the output to go to $+V_{DD}$ for B low.

NOR Gate

The circuit operation of the NOR gate, shown in Figure 6-4, is analogous to that of the NAND gate. However, the series-parallel arrangement of the MOS transistors is changed. In the NAND gate, Q_1 and Q_2 are in parallel, and Q_3 and Q_4 are in series. In the NOR gate, Q_1 and Q_2 are in series, and Q_3 and Q_4 are in parallel. Logical ORing is accomplished by paralleling transistors in both, CMOS and TTL, technologies. Logical ANDing is accomplished through the multiple emitters in TTL and by series-connected transistors in CMOS. The use of series and parallel MOS transistors is more obvious in the multiple-input CMOS gates.

As in the NAND gate, the inputs of the NOR gate are diode-protected. In addition to the smaller transistor die size, the CMOS digital device is implemented by a smaller number of components and does not include any resistors. This results in a small IC die size. Both the NAND and NOR gates may be operated as INVERTERS by connecting their A and B inputs together. In the NOR gate, the n-MOS transistors are internally paralleled. This does not change the circuit operation of the device as an inverter, but it does increase the output sink capability. Two transistors, instead of one, can now sink current. Similarly, a NAND gate with its A and B inputs connected together produces an inverter with greater source capability. This same idea is carried over to the device level, where entire gates are paralleled to increase the sink and source drive capability. The parallel devices function as circuit buffers or drivers.

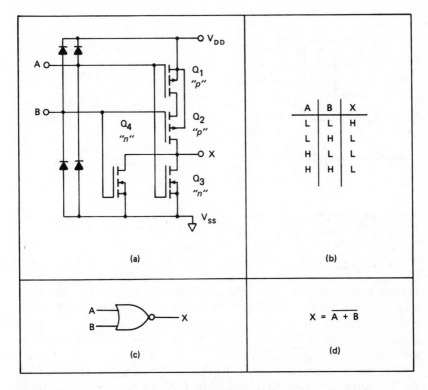

A	B	X
L	L	H
L	H	L
H	L	L
H	H	L

(a)

(b)

(c)

$$X = \overline{A + B}$$

(d)

Figure 6-4. CMOS NOR Gate. (a) Schematic; (b) truth table; (c) symbol; (d) logic equation.

Transmission Gate

The CMOS transmission gate is equivalent to a single-pole, single-throw switch and is formed by the parallel connection of p-type and n-type MOS transistors. This switch expands the versatility of CMOS circuits and is characterized as having very low resistance when closed and extremely high resistance when open. The on/off states of the transmission gate are determined by the potential difference applied to the two transistor gate leads. These two gate leads are connected to the input and output of an inverter, which provides the complementary drive signals. The combination of the transmission gate and inverter forms the complete switch and is shown in Figure 6-5.

The bulk leads of the p- and n-MOS transistors are connected to V_{DD} and V_{SS}, respectively. The switch will be closed when the input is low, and it will be open when the input is high. This arrangement of the transistors in the transmission gate is advantageous in switching applications.

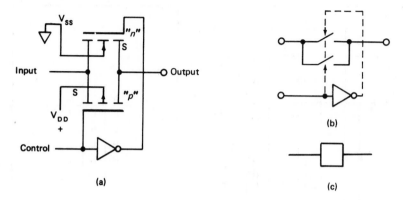

Figure 6-5. CMOS Transmission Gate. (a) Schematic; (b) functional representation.

D Flip-Flop

The block diagram of a D-type flip-flop is shown in Figure 6-6. The block diagram shows a master flip-flop comprised of two inverters and two transmission gates (shown as switches), which drives a slave flip-flop having a similar configuration. The transmission gates are driven by a clock pulse (CL) and its inverted level ($\overline{CL}$).

When the clock is at a low level, the TG_1 transmission gates are closed and the TG_2 gates are open. This configuration allows the master flip-flop to sample incoming data, and the slave to hold the data from the previous input and feed it to the output. Memory is maintained in the slave flip-flop through the feedback inverter and the closed TG_1 gate.

When the clock is high, the TG_1 transmission gates open and the TG_2 gates close, so that the master holds the data entered and feeds it to the slave. The D flip-flop is static and holds its state indefinitely if no clock pulses are applied. Figure 6-7 shows the logic diagram, symbol, and truth table for a D-type flip-flop.

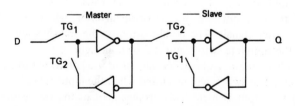

Figure 6-6. Block Diagram of a CMOS D Flip-Flop

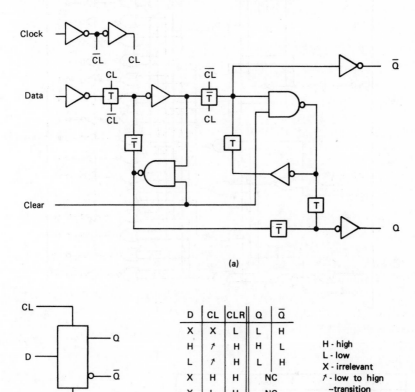

(a)

(b)

D	CL	CLR	Q	Q̄
X	X	L	L	H
H	↗	H	H	L
L	↗	H	L	H
X	H	H	NC	
X	L	H	NC	

H - high
L - low
X - irrelevant
↗ - low to high
—transition
NC - no chnge

(c)

Figure 6-7. CMOS D Flip-Flop. (a) Logic diagram; (b) symbol; (c) truth table.

6.3. MEDIUM-SCALE INTEGRATION (MSI)

Decoder

The similarity in function between TTL and CMOS continues into MSI devices. Figure 6-8 shows the CMOS equivalent of the previously described bipolar 4-to-16 line decoder. Because of the difference in technologies, the CMOS implementation (in terms of gates) varies from the bipolar version. The CMOS device uses the equivalent of 20 INVERTER, 20 NOR, and 8 NAND gates. This is a greater number of gates than the TTL decoder uses, but each CMOS gate is smaller in die size.

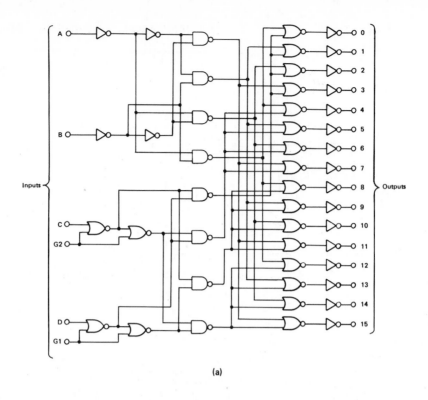

(a)

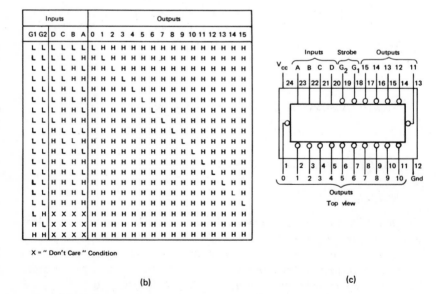

Inputs						Outputs															
G1	G2	D	C	B	A	0	1	2	3	4	5	6	7	8	9	10	11	12	13	14	15
L	L	L	L	L	L	L	H	H	H	H	H	H	H	H	H	H	H	H	H	H	H
L	L	L	L	L	H	H	L	H	H	H	H	H	H	H	H	H	H	H	H	H	H
L	L	L	L	H	L	H	H	L	H	H	H	H	H	H	H	H	H	H	H	H	H
L	L	L	L	H	H	H	H	H	L	H	H	H	H	H	H	H	H	H	H	H	H
L	L	L	H	L	L	H	H	H	H	L	H	H	H	H	H	H	H	H	H	H	H
L	L	L	H	L	H	H	H	H	H	H	L	H	H	H	H	H	H	H	H	H	H
L	L	L	H	H	L	H	H	H	H	H	H	L	H	H	H	H	H	H	H	H	H
L	L	L	H	H	H	H	H	H	H	H	H	H	L	H	H	H	H	H	H	H	H
L	L	H	L	L	L	H	H	H	H	H	H	H	H	L	H	H	H	H	H	H	H
L	L	H	L	L	H	H	H	H	H	H	H	H	H	H	L	H	H	H	H	H	H
L	L	H	L	H	L	H	H	H	H	H	H	H	H	H	H	L	H	H	H	H	H
L	L	H	L	H	H	H	H	H	H	H	H	H	H	H	H	H	L	H	H	H	H
L	L	H	H	L	L	H	H	H	H	H	H	H	H	H	H	H	H	L	H	H	H
L	L	H	H	L	H	H	H	H	H	H	H	H	H	H	H	H	H	H	L	H	H
L	L	H	H	H	L	H	H	H	H	H	H	H	H	H	H	H	H	H	H	L	H
L	L	H	H	H	H	H	H	H	H	H	H	H	H	H	H	H	H	H	H	H	L
L	H	X	X	X	X	H	H	H	H	H	H	H	H	H	H	H	H	H	H	H	H
H	L	X	X	X	X	H	H	H	H	H	H	H	H	H	H	H	H	H	H	H	H
H	H	X	X	X	X	H	H	H	H	H	H	H	H	H	H	H	H	H	H	H	H

X = " Don't Care " Condition

(b)

(c)

Figure 6-8. CMOS 4-Line to 16-Line Decoder/Demultiplexer. (a) Logic diagram; (b) truth table; (c) package.

98

The decoder accepts four binary inputs. There are 2^4 or 16 various combinations of input data, and one of the 16 outputs will go low for each combination. The device is provided with two strobe inputs (G_1 and G_2), both of which must be low for normal operation. If either strobe input is high, all 16 outputs will go high.

To use the device as a demultiplexer, one of the strobe inputs serves as a data input terminal while the other strobe is maintained low. The information at the input will then be transmitted to the selected output determined by the four-line input address. The demultiplexing function routes a single line of information to one of 16 outputs, which are selected by a four-bit address.

Shift Register

The eight-bit shift register in Figure 6-9 has gated serial inputs and eight parallel outputs and is a serial in–parallel out type of register. The flip-flops in the register are D-type master/slave flip-flops. Data are serially shifted in and out of the eight-bit register during the positive-going transition of the clock pulse. The CLEAR input is independent of the clock and will empty the register or set all outputs to a low when it is set to a low.

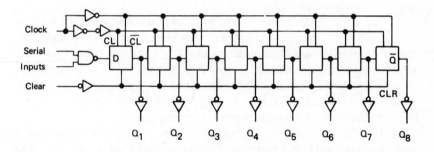

Figure 6-9. CMOS Serial in–Parallel Out Shift Register

LARGE-SCALE INTEGRATION (p, n–MOS)

6.4. RANDOM ACCESS MEMORY (RAM)

A RAM, or *R*andom *A*ccess *M*emory, is an array of memory cells where digital information is inputted, stored, and outputted by using commands and addresses. A more proper name for RAM is read/write memory, because data are accessed in the same manner for both functions. Digital data, i.e., organized 1s and 0s, may either be written into memory or read from memory repeatedly. RAM ICs are used collectively to form a storage medium in computers and digital processing machines.

Memory subsystems, implemented with a number of ICs, are generally identified by number of words, number of bits, and function. For example, a 1024 × 16 RAM is random access read/write memory containing 1024 words of 16 bits each. Memory *device* organizations follow the same rule. Individual RAM ICs are expandable to allow them to operate in conjunction with other devices and tend to be organized in *n* words by one bit to optimize lead usage.

Bipolar and MOS RAMs perform the same basic function. Bipolar RAMs are faster and are used in applications where speed is important. An example is the scratchpad memory, where interim calculation results are stored for the computer's central processor unit. An example of a MOS RAM application is a buffer memory, where data sent by the computer are temporarily stored before they are accessed by a computer peripheral.

MOS RAMs are either static or dynamic. In a static RAM, information is permanently held in a memory cell or flip-flop. In a dynamic RAM, information is temporarily held by the circuit's parasitic capacitance and must be reprogrammed or refreshed to appear permanent. The dynamic RAM requires one or more clock signals.

In the static RAM cell of Figure 6-10, two inverters are cross-wired to form a flip-flop. Devices Q_3 and Q_4 are load resistors for the Q_1 and Q_2 transistors. Devices Q_5 and Q_6 are used as two-way transmission gates and allow data to be written into the cell or read from the cell. When reading, the conducting side of the flip-flop pulls the data line toward ground by way of these gates. Writing is accomplished by forcing the data lines to the value desired in the cell, thereby overriding the contents of the cell.

A 1024 × 1-bit static RAM, with decoders, is shown in block diagram form in Figure 6-11. It uses a *n*-MOS technology and has its memory block

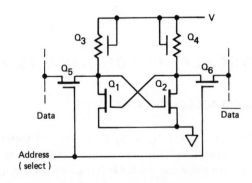

Figure 6-10. MOS RAM Memory Cell

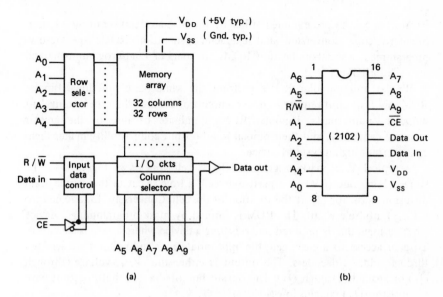

Figure 6-11. 1024 x 1 MOS RAM. (a) Block diagram; (b) device.

arranged in 32 columns and 32 rows. The rows and columns of cells are accessed through decoders whose inputs are the address lines A_0 through A_4 and A_5 through A_9. Data are read into memory through the Data In line and a given address and are read out through the Data Out line and a given address. Like many other MOS devices, the inputs and outputs are DTL/TTL-compatible. Memory may be expanded by OR-tying individual ICs to a data bus and using the chip-enable input as a control. The read or write function is determined by the level of the input signal $R/\overline{W}$.

6.5. READ ONLY MEMORY (ROM)

A ROM, or Read-Only-Memory, is a collection of preprogrammed memory cells whose digital information may be outputted by the addressing of the cells. Addressing is similar to that of a RAM, and its readout is nondestructive. A ROM offers nonvolatile storage; i.e., data are retained indefinitely even when power is shut off. ROMs are used in computers and digital processing machines, where their applications include arithmetic operations, code conversion, character and random logic generation, and microprogramming. In most cases their applications are a variation of a lookup table operation, where the ROM functions as a digital dictionary. An address input word becomes a reference to locate a new (and bigger) word. The contents of the

ROM cells can be programmed either by the IC manufacturer or by the user. Standard code conversion and character generation ROMs are factory-programmed. Special-application ROMs are user- or field-Programmable and are called PROMs.

Bipolar and unipolar ROMs perform the same basic functions. Bipolar ROMs are faster but require a greater amount of die area than their equivalent unipolar counterparts. Unipolar ROMs dominate because of the greater memory capability, and their bipolar input-output compatibility makes them an important digital systems device.

A typical MOS mask-programmed ROM cell appears in Figure 6-12. To store a logic one (0V) in a particular cell, a hole is cut in the source-drain diffusion mask for Q_3 at the location of the cell. Conversely, for a logic zero $(-V_{GG})$, no hole is cut. The ROM is formed by mask-programming a device where a logic one is required and *omitting* a device where a zero is called for. To gain access to a given cell, the appropriate row and column are selected through address decoders. The output is either the $-V_{GG}$ voltage (through Q_1) or ground (through Q_3). The output line of the cell that does not have a transistor (Q_3) remains low $(-V_{GG})$.

The widespread usage of the computer has created a need for man-machine interface or terminals. The output of these interfaces is some type of display of numbers, letters, and symbols that are called *characters*. The IC that accepts the digital system information and converts it to the proper form to drive the display is the character-generation ROM. An example of a terminal is the CRT (cathode ray tube), and its display is in the form of characters created by a dot matrix. Various characters are generated by illuminating

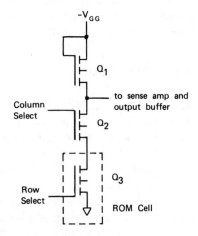

Figure 6-12. MOS ROM Memory Cell

combinations of these dots in the matrix. The 4240 is a 2560-bit *p*-MOS enhancement mode ROM programmed to generate 64 display characters in a 5 X 7 dot matrix. This device, Figure 6-13, has six address or character code inputs, five outputs to drive the dot matrix, and three row address inputs.

A character may be generated by illuminating a sequence of vertical (column) series of dots or horizontal (line) series of dots. The 4240 illuminates a character by a timed sequence of seven (row addresses) lines of 5 dots each.

The character addresses are ASCII-(American Standard Code for Information Interchange)-coded. An ASCII address (example: 100000) for a character (example: *A*) is applied to the code input-address lines of the 4240, and the row counter input address is set to 001. The number three dot is illuminated. The input address is maintained (100000), but the row address is set to 010, where the number two and number four dots are illuminated. The procedure is repeated through row address 111, where the number one and number five dots are illuminated. Although only one line of dots is lit at a time, the timing is so fast that all the lines appear to the human eye to be painted on the CRT simultaneously. The digital information is continually recycled, and the character is repainted so that it appears permanent on the CRT screen.

The operation of the ROM in a system is more complex. In actuality, a character is not painted in its entirety before the next character. Instead, all the first rows of all the characters on a CRT line are painted. Then all the

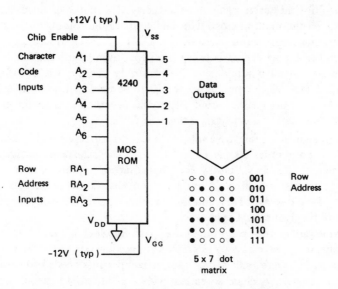

Figure 6-13. Standard Character Generator MOS ROM

second rows of all the characters are sequentially painted, until the entire CRT line is complete.

The logic surrounding the MOS ROM is bipolar to take advantage of the higher operating speed. The MOS ROM accepts bipolar input signals and generates bipolar output signals to ease the interfacing requirements. The ROM outputs typically drive a parallel in–serial out register, which drives the single CRT electron beam. The row address inputs are driven by a counter that sequences, in binary, from 000 to 111. The character code input addresses are driven by registers.

6.6. DIGITAL CLOCK

The complexity of LSI devices is exemplified by the block diagram of the digital clock IC in Figure 6-14(a). This block diagram includes the counter, decoder, multiplexer, oscillator, shaper, and ROM functions. The IC depicted is a *p*-channel MOS device.

The time base for the digital clock chip is the frequency of the line AC voltage. The standard 60-Hz AC signal is shaped and divided by six to yield 10 PPS. This signal is again divided by 10 to yield seconds, which are counted. The seconds are divided by 60 to yield hours, and again are divided by 12 or 24 before the total counting process is repeated. The outputs of the counters containing seconds, minutes, and hours information is multiplexed and decoded and serves as the inputs to a code converter or ROM. The ROM outputs drive the seven segments of a display. The multiplex divider/decoder is driven by the multiplex oscillator and controls the routing of the seconds, minutes, and hours information to the ROM's decoder. It also sequentially turns on the display digits whose segments are driven by the ROM outputs.

Various control inputs extend the versatility and application of the IC. A 50-Hz/60-Hz select input allows the IC to be used with foreign 50-Hz systems. A 12/24-hour select input allows the device to display the standard 12-hour (×2) day or the military 24-hour day.

A 4/6-digit input select permits clock manufacturers to build clocks with an hours and minutes display or an hours, minutes, and seconds display. For convenience, the clock display may be advanced rapidly, slowly, or held through the fast set, slow set, and hold controls. The *BCD* outputs are also used to drive displays. They and the seven-segment signals work in conjunction with the digit enable signals.

A wide variety of displays can be used with the clock IC. However, external components are necessary to interface the chip outputs to the display. These components are the bulk of the discrete components needed to complete a finished product. A clock schematic with a gas-discharge (neon) display is shown in Figure 6-14(b). This type of display requires higher operating voltages than the semiconductor light-emitting diode type.

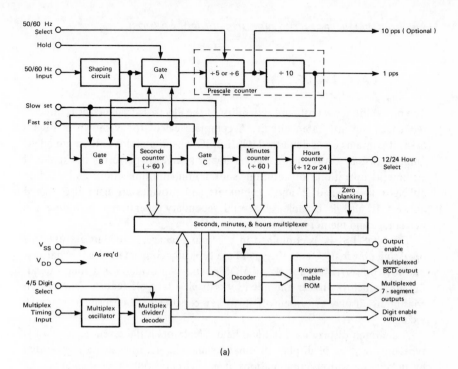

(a)

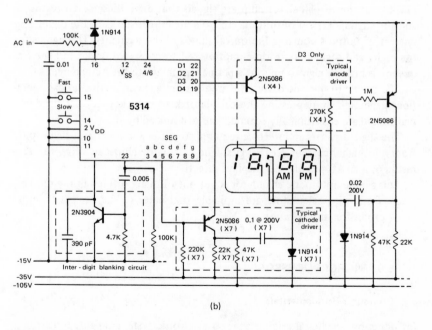

(b)

Figure 6-14. Digital Clock. (a) Block diagram; (b) application using a gas discharge display.

6.7. CALCULATOR

The portability, low cost, and reliability of the electronic calculator is directly related to the integrated circuit. A complete calculator is shown in Figure 6-15; it contains one calculator MOS/LSI IC, one digital driver IC, a nine-digit numeric display, a keyboard, and a power source. It performs the standard arithmetic operations of addition, subtraction, multiplication, and division on eight-digit signed (+ and -) numbers and provides an eight-digit signed answer. It also is capable of several secondary operations, including the percentage and memory operations.

The MOS/LSI IC is contained in a 24-pin package. Two pins are used for operating power, V_{SS} and V_{DD}, and are typically connected to a 9-V battery. The low power consumption of MOS circuitry provides for a long operating life, which is further extended by a display turn-off feature. The power-consuming display is turned off if no key closures have been made after 16 seconds.

A common display used in hand-held calculators is the seven-segment LED type. In this type of display all numbers and the negative sign are generated by turning on various combinations of the light-emitting diode segments. The outputs of the IC calculator chip are signals that drive these seven segments and the digits. One set of outputs controls the on/off status of the digits, and one set of outputs controls the on/off status of the segments. The signals for the digits and segments are time-multiplexed to conserve power, but the frequency is high enough to present a permanent image to the human eye.

The inputs to the calculator chip are the numbers and mathematical operations generated by depressing the calculator keys. The keys are arranged in a matrix and are electronically scanned and debounced by the IC.

The digit driver is an interface required to buffer the calculator IC outputs from the higher-current LED digits. The calculator chip can drive the segments of many types of LED displays directly.

This simple arithmetic calculator is but one of many offered in the marketplace. Currently, calculators are available that cater to the needs of almost every profession and vocation.

6.8. MICROPROCESSOR

Computer Fundamentals

Introduction. A digital computer is an electronic system capable of carrying out repetitive and highly complex mathematical, logical, and control operations at high speed. The digital computer is comprised of digital circuits.

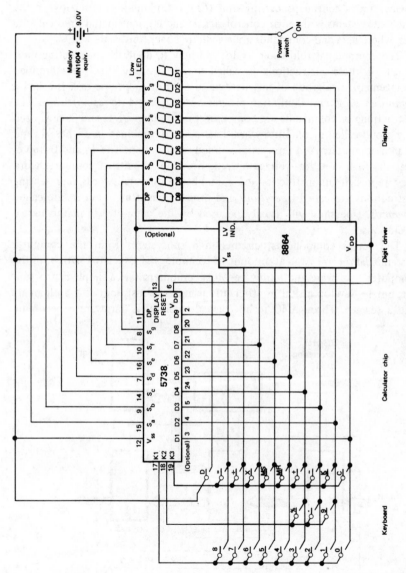

Figure 6-15. Calculator

The block diagram of a digital computer is shown in Figure 6-16. The computer is comprised of three subsystems: memory, an input-output (I/O) device, and a central processing unit (CPU). Communications between the three subsystems is through control signals and a set of digital lines called a *bus*, where data and commands are transferred back and forth.

The computer is told what to do, and how to do it, through instructions. A set of these instructions is called a *language*, and the predetermined arrangement of a group of instructions to perform a specific task is called a *computer program*. Computer programs are referred to as *software*. The instructions or commands are composed of a number of 1s and 0s organized in a group called a computer word. Each unique combination of these 1s and 0s in a computer word represents a unique instruction. The 1s and 0s represent the high- and low-voltage outputs of digital circuits. A computer word may represent an instruction or a number. Computer words representing instructions are called *operations* and those representing numbers are called *operands*. Operands will usually be used by the computer in mathematical operations.

The actual components, circuits, and subsystems form the computer system's *hardware*. A great amount of technology is associated with both the computer's software and hardware. Most computers have a simple instruction set, but to develop a skill in efficiently using these instructions is both an art and a science. Through IC technology, the computer hardware is becoming

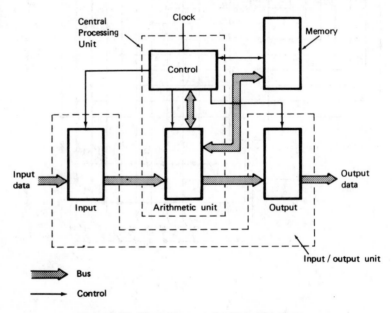

Figure 6-16. Block Diagram of a Digital Computer

smaller and smaller in size, but the *detailed* understanding of the functioning of its operation is no easy task.

Memory. The computer contains a memory (flip-flops) subsystem that stores or memorizes the instructions and operands. A flip-flop or memory cell stores one bit, and a group of cells stores the computer word that is representative of an instruction or number. The instructions and operands are stored in memory in the order that they are to be performed or operated on. The computer, initially, is forced to go to the first instruction and is automatically indexed to the next when it is completed. The procedure is repeated until the program is terminated and the task completed. The instructions and operands are stored in specific memory locations determined by a memory address.

Central Processing Unit. The computer subsystem that performs the control, timing, and temporary storage of mathematical and logical operations, and controls the inputting and outputting of data is called the CPU or Central Processing Unit. It is the brains and heart of the computer. A control and Arithmetic Logic Unit (ALU) make up the CPU. The ALU contains adders and shift registers used in arithmetic operations and registers for temporary storage. The control unit contains decoders and combinational (gates) logic for control and timing, and ROMs that are used for control and microinstruction (not programmable) sets.

Input/Output Device. Communications between man and the computer are accomplished through the Input-Output or I/O subsystem. Examples of I/O devices include the teleprinter, CRT, and paper tape reader and punch. Information in the form of instructions and data is usually entered in the computer from a typewriter-type keyboard associated with the CRT and teleprinter units. Information is outputted from the computer to the printer section of these same I/O units. The keyboard and printer characters, i.e., the letters, numbers, and symbols, are converted to 1s and 0s by the I/O device. The paper tape reader and punch use holes or no-holes in paper as their input and output medium.

An Integrated Microprocessor

The microprocessor (uP) is a general-purpose computer whose central processing unit (CPU) is on a set of integrated circuit chips. Currently, several manufacturers have condensed the CPU to one IC. The microprocessor has evolved from the architecture of the advanced minicomputer systems using the technology employed in producing advanced calculators.

The 8080 is a complete eight-bit parallel central processing unit for use in general-purpose digital computer systems. It is fabricated on a single LSI chip by using an n-channel MOS process and is the most popular microprocessor produced. A complete microprocessor *system* is formed when the 8080 CPU is interfaced to an I/O unit and any type of semiconductor memory. The device is contained in a 40-pin package, Figure 6-17, and its operation is depicted in a simplified functional block diagram.

The CPU data bus in the functional block diagram of Figure 6-17(b) is eight digital lines that provide bidirectional communication between the CPU, memory, and I/O devices. Data and instructions are passed on these lines, D_0 through D_7, and temporarily stored in the data bus buffer/latch. The information is then routed to various CPU locations through an internal eight-bit data bus that is not externally accessible. A second set of digital lines, A_0 through A_{15}, is used to address memory or the I/O devices. This address bus identifies the locations where information is coming from or going to.

Incoming instructions or machine directives are temporarily held in an instruction register and decoded to provide the timing and control information for the CPU's next operation. Information may be temporarily stored in one of several registers (W through L) prior to its being operated on, or before the information is outputted. Two registers, the stack pointer and program counter, are dedicated to specific tasks. The program counter always points to the location of the CPU's next instruction and is automatically incremented after the completion of the previous instruction. This register is automatically and sequentially indexed through the instructions and data that the CPU must operate on. The stack pointer is a register that performs a similar function on a data table or list of operands.

Arithmetic operations are performed by the arithmetic logic unit (ALU). Associated with the ALU are registers and accumulators that temporarily store multiple operands or intermediate results. Data are inputted and outputted via the internal bus through directions provided by the timing and control unit.

In addition to the bus and address lines, the 8080 has twelve control signals. The synchronous operation of the CPU is controlled by an external clock through the signals ϕ_1 and ϕ_2. The data bus-in or DBIN signal indicates to external circuits that the data bus is in the input mode. This signal is used to enable the gating of data onto the 8080 data bus from memory or I/O. The write or $\overline{\text{WR}}$ signal is used for memory write or I/O output control. Because of the high-speed capability of the CPU, several I/O devices can be operated at one time. These I/O devices request the processor to service them through interrupt control lines. The interrupt request signal, INT, and the interrupt enable, INTE, control this operation. WAIT, RESET, HOLD, SYNC, and hold acknowledge, HLDA, complete the CPU control signals.

The CPU chip is powered with +12, +5, and –5 V DC. It has one DC reference or ground pin.

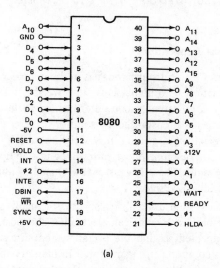

(a)

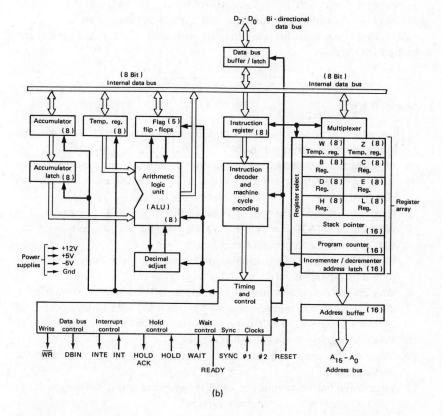

(b)

Figure 6-17. 8080 Microprocessor. (a) Pin configuration; (b) CUP functional block diagram.

Microprocessor Applications

The greatest impact of electronics today is in the microprocessor applications area. The combination of minimal cost, low weight, small size, low energy conversion, and the power of a computer has led all facets of business and industry to look to it for cost-effective solutions. Presently, microprocessors can be found in electronic cash registers, weighing scales, video games, traffic light directors, industrial controllers, automated test equipment, and aviation and marine subsystems. These are just a few examples illustrating the microprocessor applications diversity.

The design of a system using a microprocessor involves the interfacing of major functional blocks. Mechanical, electronic, chemical, and optic subsystems are interconnected so that the overall system can perform a high-level task. Typically, a computer program stored in memory interprets data from system sensors or transducers, and based on this data initiates the proper sequence of commands in response to its inputs. As an example, a microprocessor in the traffic light director will control the red, green, and yellow traffic signals based on the rate of traffic derived from sensors located at the traffic light location.

QUESTIONS

1. Identify the three basic types of FET transistors. Which ones are normally found in integrated form?
2. Why are the p-MOS devices in the NAND gate of Figure 6-3 connected to V_{DD}? Can the positions of the p- and n-MOS transistors be exchanged? Why?
3. How is the fanout or output drive capability of CMOS gates increased?
4. Why are the MOS transistors paralleled in the switch portion of the CMOS transmission gate shown in Figure 6-5?
5. Why are unipolar ROMs capable of having a greater amount of stored data than bipolar ROMs?
6. Why are the inputs and outputs of unipolar ROMs bipolar compatible?
7. What are the two primary reasons for using MOS transistors instead of bipolar transistors in a calculator integrated circuit?
8. List the functions found in a digital clock IC.
9. How is the operating life of a battery in a hand-held calculator maximized?
10. What three hardware modules must be added to a microprocessor IC to make a complete microprocessor *system*?

PROBLEMS

1. Draw the schematic and logic diagram for a CMOS AND-OR-INVERT gate.
2. How many outputs will a decoder have if it had six inputs and it was similar to the one of Figure 6-8?
3. What is the minimum number of row and column selector inputs required for a 16, 384 × 1 MOS RAM?
4. How many segments are on if a 9 is illuminated when a semiconductor seven-segment display is used? How many segments are on if a 2 is illuminated?
5. What changes in the digital clock block diagram of Figure 6-14(a) are required if the clock must operate in a 40-Hz system?

Linear Integrated Circuits

Amplifiers

Process, control, signal generation, and amplification operations are performed by linear circuits. This chapter discusses linear integrated circuits that amplify.

7.1. INTRODUCTION

A linear integrated circuit is a configuration of electronic components manufactured on a continuous piece of material whose inputs and outputs are *proportionally* or *mathematically* related for all values within the operating range of the circuit.

In digital circuits, two values of voltage contain information. Typically a voltage very near 0V represents a logic 0 and a voltage near +5 V (or V_{CC}) represents a logic 1. The high and low voltage or logic 1 and 0 represents a TRUE and NOT TRUE state of a digital logic function. In linear circuits *all* values of voltages contain information, and the number of values is unlimited and continuous.

The typical linear circuit has one output whose reference is the circuit's common or ground, and it may have zero, one, or several inputs. In digital circuits, the input and output variables are pulsed DC voltages. In linear circuits, the inputs and outputs are pulsed DC, DC, or AC voltages. In fact, the input and output voltages do not have to be of the same type. Although generally the input-output variables are voltage, other electronic variables are not uncommon in linear circuits, including current, capacitance, resistance, inductance, and frequency.

OPERATIONAL AMPLIFIER

7.2. OPERATIONAL AMPLIFIER FUNDAMENTALS

To amplify means to make larger. The requirement to amplify electronic variables is extensive and not necessarily limited to voltage and current. Linear circuits are used to amplify power, capacitance, inductance, resistance, and even frequency. However, the greatest usage of amplifiers is for AC and DC voltages. Voltage amplifiers may be classed in two categories: (1) low-frequency amplifiers, and (2) high-frequency amplifiers. The amplification of high-frequency signals plays an important role in TV, radio, and communications systems. Low-frequency amplifiers are applicable where signal frequencies vary from DC to about 1 Megahertz.

The low-frequency amplifier is the monolithic operational amplifier or op amp. It is the most popular and widely used linear IC in the world and is sold by the tens of millions of devices per year. The op amp can be found in diverse applications and derives its name from those amplifier circuits that perform mathematical *operations*.

The operational amplifier is a two-input, one-output *voltage amplifying* linear device that is capable of high gain, high input resistance, and low output resistance. Table 7-1 lists the numbers associated with these characteristics for five design generations of monolithic amplifiers.

TABLE 7-1

Amplifier					
Characteristic	709C	741C	301A	308	156
Voltage gain	45,000	160,000	160,000	300,000	200,000
Input resistance	250KΩ	1MΩ	2MΩ	40MΩ	10^6MΩ
Output resistance	150Ω	75Ω	75Ω	75Ω	75Ω

Typical specifications

The op amp schematic symbol, the first order circuit and mathematical models, and its transfer characteristic are shown in Figure 7-1.

The amplifier is a differential amplifier and has two input terminals designated noninverting (+) and inverting (−). The differential inputs allow the amplifier to magnify the difference of two voltages. The input-output voltage relationship is given as

$$V_{out} = -A\,V_{in}$$

where A is the voltage gain of the amplifier and V_{in} is the potential difference between the inputs. The output, a single terminal, is referenced to the

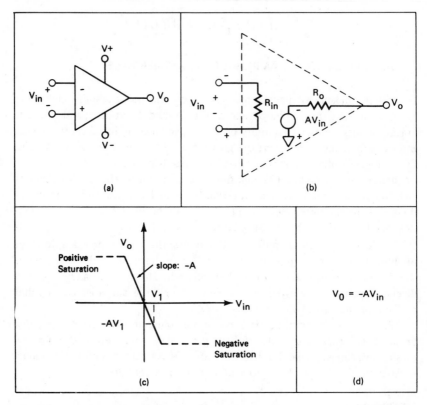

Figure 7-1. IC Operational Amplifier. (a) Schematic symbol; (b) circuit model; (c) Input/output transfer characteristic; (d) mathematical model.

power supply ground or common measurement point and is capable of swinging to a positive or negative voltage.

If the input voltage to the amplifier is such that the inverting input (−) is positive with respect to the noninverting terminal (+), the output will be a negative voltage with respect to ground. The magnitude of the output will be A times the difference of the inputs. If the noninverting (+) input is positive with respect to the inverting input (−), the output voltage will be positive with respect to ground. The magnitude of the output voltage will be A times the difference of the inputs.

Some op amps have single inputs and some have differential outputs, but they are few in number. In addition to the input and output terminals, the amplifier must have two power supply pins that are labeled V^+ and V^-. Typically, the amplifier is powered with ±5 V DC to ±20 V DC. Additional pins, usually three, are provided to allow the user to compensate for inherent amplifier deficiencies or to control its frequency response and bandwidth. The op amp contains more than just gain-producing circuits. Included in the design are features that make its application simple and reliable. Some of these features are input and output overload protection, no latch-up or oscillations, and simple frequency response compensation.

The transfer characteristic (Figure 7-1c) graphically displays the input to output voltage relationship. For any value of input voltage, the output is A times that value. The polarity of the output is a function of the polarity of the applied input voltage relative to the amplifier's input terminals. For the transfer characteristic, the more positive input voltage is applied to the inverting (-) input. The slope of the diagonal line is $-A$ and illustrates the proportional or linear relationship of the input and output. This linear relationship is maintained until the amplifier output is voltage-limited. The limit in the positive direction is called *positive saturation* and in the negative direction is called *negative saturation*. Saturation refers to the operating state of the transistors of the amplifier's output stage at these limits.

In the model for the operational amplifier, the source, AV_{in}, is an ideal, dependent voltage source. Its value depends on the value of V_{in}, where V_{in} is the differential input voltage. The resistance between the two input terminals is R_{in}, and the resistance in series with the output is R_0.

The operational amplifier, generally speaking, is a low-frequency amplifier. It has a high voltage gain (200,000) near DC, but the gain decreases as frequency increases beyond the amplifier's cutoff frequency, which is at 5–50 Hz. The gain decreases, by 20 dB, for every decade increase in frequency until the open loop amplifier gain is 1 at about 1 MHz. The amplifier is not used for signals whose frequencies are above this point. The gain versus frequency amplifier characteristic is illustrated in the Bode plot of Figure 7-2.

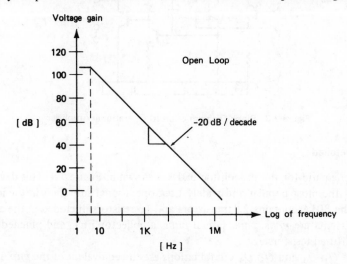

Figure 7-2. Gain versus Frequency Characteristic of the Operational Amplifier

The Bode plot is an idealized graph of the voltage gain of the device, expressed in dB, versus the log of frequency. For low-frequency applications where additional frequency stability is required, the curve may be shifted to the left by increasing the value of the compensation capacitor.

7.3. INTEGRATED OPERATIONAL AMPLIFIER

The block diagram for a typical op amp is shown in Figure 7-3. It has three basic stages. The first stage consists of a differential amplifier whose function is to amplify the difference between two input voltages. This stage is characterized by two *pnp* transistors with a common emitter. The second stage consists of voltage-amplifying, frequency compensation, and DC level shifting circuits. It is characterized by a common-emitter transistor with an unbypassed emitter resistor. The third or output stage is the workhorse of the amplifier. Its output must swing positive and negative with respect to ground and deliver the relatively heavy output current. Its output resistance must be low, and the stage must be protected against adverse loading conditions. It is characterized by the class AB output stage implemented with an *npn* and *pnp* transistor.

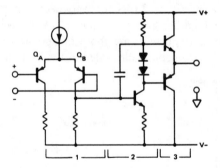

Figure 7-3. Block Diagram of an IC Operational Amplifier

Detailed

The schematic for the monolithic 301A is shown in Figure 7-4. This device is one of the most popular and widely used operational amplifiers in the industry. The 301A contains 22 transistors and 16 resistors. Included in the above components are *npn*s, *pnp*s, lateral *pnp*s, a collector FET, and pinched base and diffused resistors.

The Q_1-Q_3 and Q_2-Q_4 combinations are the equivalent of the *pnp* differential amplifier (Q_A-Q_B) shown in the block diagram. This stage must

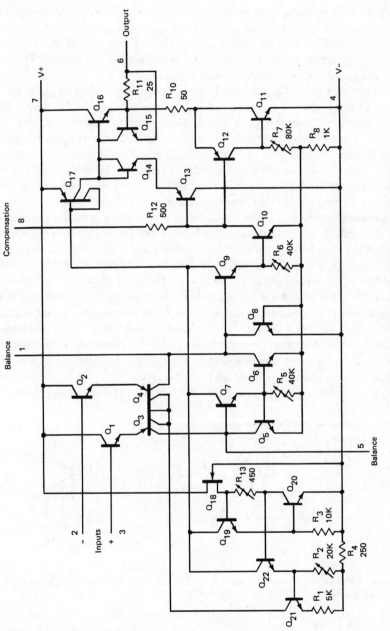

Figure 7-4. Schematic of the 301A IC Operational Amplifier

119

amplify voltage differences for all values of input voltages (common-mode), must have a high input resistance, a low offset voltage error (ΔV_{BE}), and have a small input current (base current). The collector load resistors for the differential amplifier are *simulated* with active devices (transistors). Q_5 and Q_6 and associated circuitry form these active loads.

Transistors Q_9 and Q_{10} form a Darlington common emitter amplifier stage with an unbypassed emitter resistor (R_8). The active load for this stage is the Q_{17} circuit. Transistor Q_{17} is one of three multiple-collector *pnps* in the 301A. These devices are made by splitting the collector (Figure 7-5) of a lateral *pnp* transistor into two or more segments. By connecting one segment back to the base, the current gain is then determined by the relative size of the segments. This circuit/device technique is extensively used in linear IC designs to provide the less sensitive current-source circuit bias. Frequency response compensation is accomplished by connecting a capacitor (internally or externally) from the collector to base of the Darlington (Q_9-Q_{10}) pair. Externally this is done through device pins 1 and 8 of a standard eight-pin T05 package. Pin 1 is also used in conjunction with device pin 5 to correct for the offset voltage error in the input differential amplifier. This error is primarily caused by the mismatch in the base-to-emitter voltages of the two amplifier input transistors. In terms of amplifier performance, offset voltage means that at zero volts input, the output voltage will *not* be zero. In the transfer characteristic [refer to Figure 7-1(c)], this input offset voltage (V_{OS}) shifts the diagonal line slightly to the left or right. Typically, the offset voltage is less than 5 mV. The offset correction is made by unbalancing the differential amplifier active loads, which causes the *B-E* voltages to shift.

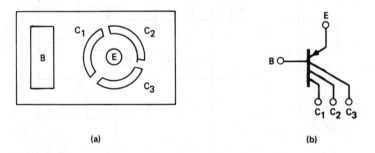

(a)   (b)

Figure 7-5. Multiple Collector Lateral pnp. (a) Top view of integrated structure; (b) schematic symbol.

Transistors Q_{16} and Q_{12} and associated circuitry form a class AB output stage. This type of output stage uses separate transistors to provide the two polarities of output current; however, both output transistors are on near zero volts output. For most of the output voltage range, one or the other transistor is off. Q_{15} and R_{11} form a current-limiting circuit. The output current flows through R_{11}. When this current begins to exceed 20 mA, the voltage drop across R_{11} approaches 500 mV, which begins to turn the base-to-emitter junction of Q_{15} on. As Q_{15} turns on, it shunts the base current of the output driver Q_{16}. As Q_{16} turns off, current limiting occurs. The current available at the base of Q_{16}, which is shunted to the output load, is fixed.

7.4. BASIC AMPLIFIER APPLICATIONS

High Gain Implies Small Input Voltage

The op amp has high gain, where "high" translates to approximately 200,000. The maximum output voltage from the amplifier is typically ±15 V. If you divide the maximum output voltage by the gain, you get the maximum input voltage required, which is ±75 μV. Signals at this level are extremely difficult to process because of noise and interference. In addition, gains of 200,000 are seldom required. Why, then, is the op amp so popular and how is it really used?

Stability Through Feedback

Op amps are used primarily in *circuits* that contain feedback, usually negative feedback. Feedback, generally speaking, implies a situation in which output quantities in a system are allowed to affect the input quantities that caused them in the first place. Negative feedback is applicable when the output signal is purposely allowed to *reduce* the effect of the input signal. This, of course, results in the reduction of the net gain of the circuit. Feedback is introduced in an actual circuit by connecting a component, usually a resistor, from the output back to the inverting (-) input.

What do we receive and what do we lose in amplifier feedback circuits? We lose the maximum obtainable gain, which would be the gain of the amplifier itself. We receive stability, primarily gain stability. Most applications do not require high gain, but they do require the gain to be precise and stable. On an op amp data sheet, the amplifier gain is guaranteed to be a certain minimum amount but the maximum is not specified, and for a given amplifier type, it will vary 10 to 20 percent. In addition to this relatively large variation, the gain of any specific amplifier will vary with temperature, age, and operating conditions. In fact, temperature is a crucial factor in all integrated circuits. Bipolar IC's use bipolar transistors whose base-to-emitter voltage is temperature dependent (-2 mV/°C) and whose current gain β is also temperature dependent. These two factors directly contribute to amplifier gain variations. A feedback configuration reduces these effects by several orders of magnitude. Besides gain stability, negative feedback also tends to improve linearity, to reduce output resistance, and, in some configurations, to increase input resistance. If the amplifier has a gain of 200,000 and we need a circuit gain of only 20, then the price paid for the advantages received is indeed modest. For the excess gain that we do not require we receive gain stability. Most voltage amplification requirements vary from 0.1 to 50. The number 200,000 is referred to as the *open-loop gain* of the amplifier. If we configure the amplifier in a circuit employing negative feedback so that the circuit's gain is 20, then 20 is called the *closed-loop gain*. Similarly, we can talk about open- and closed-loop input resistance, output resistance, and bandwidth.

The application of negative feedback to an operational amplifier yields an amplifier circuit with the closed-loop characteristics determined primarily by the values of the discrete feedback components whose accuracy and stability are easier to control.

The Ideal Amplifier

The easiest way to analyze op amp circuits is to treat the amplifier as "ideal." The ideal amplifier has infinite gain, infinite input resistance, infinite bandwidth, and zero output resistance. Here, the term *resistance* is used for simplicity. *Impedance* is the more proper term to use, since there are distributed reactive components associated with the input and output, but their role is small in the typical amplifier frequency range. These idealized properties allow us to say:

1. *The amplifier differential input voltage is zero.* This is not a bad assumption, since the real amplifier is required to have a maximum of $\pm 75\,\mu V$ and circuit voltages are in the high mV and V regions. This statement means that both amplifier terminals must *track* each other. If a voltage is applied to the noninverting input, the amplifier will drive the inverting input, via the feedback components, to the same value. If one input is at a fixed voltage or ground, the other input will be very near that value.
2. *The current entering or leaving the amplifier inputs is zero.* This is not a bad assumption either, since the input currents are typically less than 100 nA and circuit currents are in the high μA region or greater. In a real amplifier these input currents are called *bias currents*, and they are the base currents of high-gain transistors in the input differential amplifier stage.

Inverting Amplifier

The noninverting terminal of the amplifier in the inverting circuit in Figure 7-6(b) is grounded or at 0 V. The inverting terminal must track the noninverting and hence it will be electrically zero volts (not ground). The voltage across R_1 will be V_s volts, and hence a current will flow through R_1 of value V_s/R_1.

This current cannot enter the inverting input because of point 2 and must be sunk by the amplifier output via R_2. The output will go to a voltage of $-IR_2$, or

$$V_0 = -IR_2 = \left[\frac{V_s}{R_1}\right] R_2 = \frac{-R_2}{R_1} V_s$$

The closed-loop gain of the circuit is determined by the ratio $-R_2/R_1$. Since gain is the ratio of the output over the input, then

$$A_{CL} = \text{closed-loop gain} = -R_2/R_1$$

It can be less than one $(R_1 > R_2)$ or greater than one $(R_1 < R_2)$. To achieve reasonable accuracy and stability, the closed-loop gain is usually made 500 times less than the amplifier open-loop gain. This restricts the circuit closed-loop gain to a practical maximum of 400 at DC. The input resistance for the *circuit* will be R_1. The output resistance of the circuit will be significantly smaller than the output resistance of the amplifier.

The circuit can be used to

1. Provide voltage gain.
2. Generate a signal of the opposite voltage polarity.
3. Generate an output signal 180 degrees out of phase with respect to its input.

Noninverting Amplifier

The noninverting terminal of the amplifier in the noninverting circuit of Figure 7-6(a) is connected to the voltage source V_s. The amplifier output will drive the inverting input to the same value or

$$V_O = +V_s$$

The + and − signs in the defining equations indicate whether the output voltage is the same or of the opposite polarity as the input. Although the circuit neither provides gain nor performs a mathematical operation, it is valuable as a buffering device. It has a high input resistance and a low output resistance, which are two qualities necessary to interface two circuits whose input and output resistances could adversely affect each other.

Noninverting Amplifier with Gain

Both of the previous circuits are fundamental in nature and can be extended to perform more advanced functions. Figure 7-7 shows the basic buffer amplifier with gain capability added. V_s, an ideal voltage source, is connected to the (+) terminal of the amplifier. The output will drive the R_1-R_2 divider until the (−) terminal of the amplifier is at a value of V_s volts. The voltage

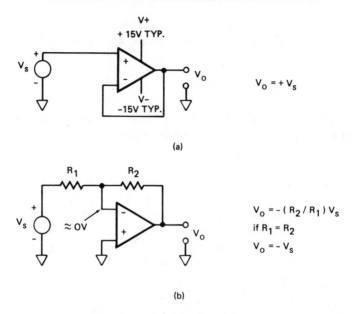

$$V_o = + V_s$$

(a)

$$V_o = - (R_2 / R_1) V_s$$
$$\text{if } R_1 = R_2$$
$$V_o = - V_s$$

(b)

Figure 7-6. Fundamental Amplifier Circuits. (a) Noninverting; (b) inverting.

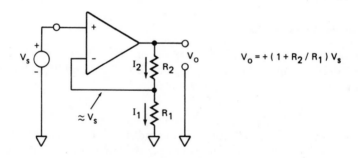

$$V_o = + (1 + R_2 / R_1) V_s$$

Figure 7-7. Non-Inverting Amplifier Circuit with Gain

drop across R_1 will be V_s, and the current flowing through it will be V_s/R_1 amperes. This current can come only from the output. Thus I_2 will be equal to I_1 and

$$V_O = V_s + I_2 R_2 = V_s + \frac{V_s}{R_1} R_2 = V_s \left(1 + \frac{R_2}{R_1}\right)$$

For this circuit the closed-loop gain

$$A_{CL} = 1 + R_2/R_1$$

The circuit can be used to

1. Provide voltage gain.
2. Buffer or interface two circuits.

Example

If $V_S = -3$ V, $R_1 = 2$ kΩ, and $R_2 = 6$ kΩ, then the gain

$$A_{CL} = \frac{V_o}{V_s} = 1 + 6 \text{ k}\Omega/2 \text{ k}\Omega = 1 + 3 = 4$$

and the output voltage

$$V_O = 4V_s = 4 \ (-3 \text{ V})$$

$$V_O = -12 \text{ V}$$

Summing Amplifier

The algebraic operation of summing means the addition of signed (+ and -) numbers. Figure 7-8 shows a *summing* amplifier circuit. Its behavior is an extension of the inverting circuit. The voltages across R_1, R_2, and R_3 will be V_{S1}, V_{S2}, and V_{S3}, because the inverting input of the amplifier is at zero volts. The inverting input for this configuration is called the "summing" junction, since it will sum the currents I_1, I_2, and I_3. These currents will flow through R_4 to the amplifier output. The amplifier output voltage

$$V_O = -R_4(I_1 + I_2 + I_3) = -\left(\frac{V_{S1}}{R_1} R_4 + \frac{V_{S2}}{R_2} R_4 + \frac{V_{S3}}{R_3} R_4\right)$$

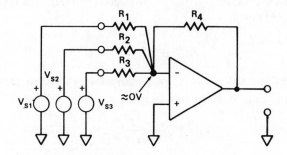

$$V_o = -[\ (R_4 / R_1) \ V_{S1} + (R_4 / R_2) \ V_{S2} + (R_4 / R_3) \ V_{S3} \ ]$$

if $R_1 = R_2 = R_3 = R_4$

$$V_o = -[\ V_{S1} + V_{S2} + V_{S3} \ ]$$

Figure 7-8. Summing Amplifier Circuit

The gain for each input voltage can be different. However, for $R_1 = R_2 = R_3$, the closed-loop gain

$$A_{CL} = -\frac{R_4}{R_1}$$

and if $R_1 = R_2 = R_3 = R_4$, then

$$V_O = -(V_{S1} + V_{S2} + V_{S3})$$

which more explicitly shows the summing operation. The circuit can be used to

1. Sum signed voltages.
2. Amplify the sum of signed voltages.

Example

If $R_1 = R_2 = R_3 = 5$ kΩ, $R_4 = 50$ kΩ, $V_{S1} = -5$ V, $V_{S2} = +6$ V, and $V_{S3} = -2$ V, then the gain

$$A_{CL} = -\frac{50 \text{ k}\Omega}{5 \text{ k}\Omega} = -5$$

and the output voltage

$$V_O = -5(-5 \text{ V} + 6 \text{ V} -2 \text{ V}) = -5(-1 \text{ V})$$
$$V_O = +5 \text{ V}$$

Difference Amplifier

The *difference* amplifier, Figure 7-9, is the complement of the summing amplifier and allows the subtraction of two signed voltages. The voltage at the noninverting (+) input

$$V_{NI} = \frac{R_2}{R_1 + R_2} \cdot V_{S2}$$

Since the two amplifier inputs track each other with their net difference ideally equal to zero, then the voltage at the inverting (−) input

$$V_I = V_{NI} = \frac{R_2}{R_1 + R_2} V_{S2}$$

The current that flows through R_1,

$$I_1 = \frac{V_I - V_{S1}}{R_1} = \left(\frac{R_2}{R_1 + R_2} \frac{V_{S2}}{R_1} - \frac{V_{S1}}{R_1} \right)$$

Since no current enters the amplifier inputs,

$$I_1 = I_2$$

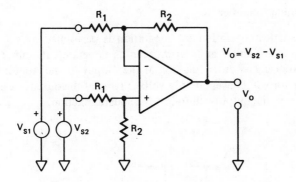

Figure 7-9. Difference Amplifier Circuit

and the output voltage will be the sum of the voltage at the (-) input plus the product of I_2 and R_2.

$$V_O = \frac{R_2}{R_1 + R_2} V_{S2} + \left[\left(\frac{R_2}{R_1 + R_2} \right) \cdot \frac{V_{S2}}{R_1} - \frac{V_{S1}}{R_1} \right] R_2$$

$$V_O = \frac{R_2}{R_1} \left(V_{S2} - V_{S1} \right)$$

The closed-loop gain

$$A_{CL} = R_2/R_1$$

$$\text{If } R_2 = R_1, A_{CL} = 1 \quad \text{then} \quad V_O = V_{S2} - V_{S1}$$

which more explicitly shows the difference operation.

The circuit can be used to

1. Find the difference between two voltages.
2. Amplify the difference of two voltages.
3. Convert a potential difference to a voltage whose reference is ground.

Example

If $R_1 = R_2 = 2 \text{ k}\Omega$, $V_{S1} = -4$ V and $V_{S2} = -8$ V, then the gain

$$A_{CL} = 1$$

and the output voltage

$$V_O = 1 \left[-8 \text{ V} - (-4 \text{ V}) \right] = 1 (-8 \text{ V} + 4 \text{ V})$$

$$V_O = -4 \text{ V}$$

Integrating Amplifier

The mathematical operation of integration is performed by the circuit of Figure 7-10 and is called an *integrator*. In an integrator, the input voltage is converted to a higher-order function at the output. If the input is a DC voltage, the circuit will convert it to a voltage ramp at the output. If the input is a voltage ramp, the output will follow the square function.

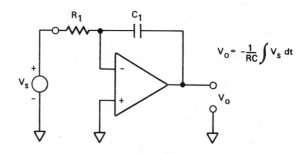

$$V_o = -\frac{1}{RC} \int V_s \, dt$$

Figure 7-10. Integrating Amplifier Circuit

Mathematically,

$$k_1 \text{ is converted to } k_2 t, \text{ or}$$

$$k_1 t \text{ is converted to } k_2 t^2, \text{ or}$$

$$k_1 t^2 \text{ is converted to } k_2 t^3, \text{ etc.}$$

where k_x are constants and t is time. The constant k_1 is analogous to a DC voltage. The function $k_2 t$ is analogous to a voltage ramp; i.e., the value of voltage increases linearly with time.

If the circuit input V_s is a DC voltage, i.e., a constant with respect to time, then

$$I_1 = V_S/R_1$$

This DC current will charge the capacitor C_1. For a positive V_s, the output voltage will be negative, since the capacitor terminal at the inverting (–) input is at zero volts and the capacitor is being charged plus (inverting input) to minus (output).

If at $t = 0$ the output voltage is zero, then

$$V_o = -\frac{V_s}{R_1 C_1} t$$

where the slope of the line or ramp

$$m = -\frac{V_s}{R_1 C_1}$$

The circuit can be used to

1. Integrate an input voltage.
2. Provide a voltage ramp for timing and sweep circuits.

Example

If $R_1 = 1$ kΩ, $C_1 = 1$ μF, and V_s is switched from 0V to +1 V at t_o, then the output will go from zero volts (t_o) in a negative direction with a slope

$$m = -\frac{1 \text{ V}}{(1 \text{ k}\Omega)(1 \text{ } \mu\text{F})} = -1 \text{ V} / 1 \text{ msec}$$

the current charging the capacitor

$$I_1 = \frac{1 \text{ V}}{1 \text{ k}\Omega} = 1 \text{ mA}$$

Current-to-Voltage Converter

The output variable for many devices, sensors, and transducers is current. This current can be converted to a voltage with the *current-to-voltage converter* circuit of Figure 7-11. The current from the source I_s flows through the feedback resistor R_1 because, ideally, the amplifier inputs can neither source nor sink any current. The amplifier output voltage

$$V_O = -I_s R_1$$

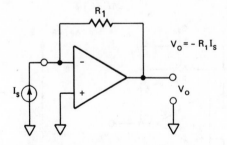

Figure 7-11. Current to Voltage Convertor

The current source output is voltage-clamped to zero volts because the inverting (-) input tracks the noninverting (+) input, which is at ground potential.

The circuit can be used to

1. Convert current to voltage.
2. Amplify photocell, photodiode, and photovoltaic cell current signals.

Example
If R_1 = 10 MΩ and I_s = 100 nA, then

$$V_O = -(10^7 \ \Omega)(10^{-7} \ A)$$
$$V_O = -1 \ V$$

Low-pass Filter

A *filter* is a circuit that exhibits a preference for signals of a certain range of frequencies over others that are outside this range.

The addition of a capacitor in parallel with the feedback resistor in the basic inverting amplifier produces a low-pass filter. The filter, Figure 7-12(a), will ideally amplify all signal frequencies below a certain frequency, called the *corner* or *break frequency*, with a gain of $-R_2/R_1$. For signal frequencies greater than the corner frequency, the amplitude of the signals decreases at a rate of -20 dB per decade. The Bode plot of Figure 7-12(b) displays the circuit's gain versus frequency characteristic. This can also be intuitively understood by considering the reactance of the capacitor. For very low frequencies, the reactance is extremely high, the feedback component is essentially the resistor R_2, and the circuit gain is $-R_2/R_1$. For very high frequencies, the capacitive reactance is small, the feedback component is essentially the capacitor C_1, and the circuit gain is low. This circuit passes low-frequency signals and attenuates high-frequency signals. Its corner frequency,

$$f_c = \frac{1}{2\pi R_2 C_1}$$

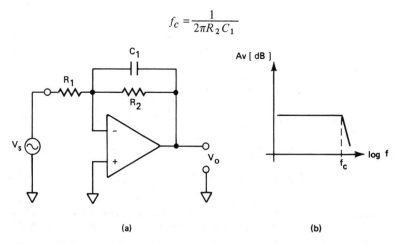

(a)   (b)

Figure 7-12. Low-Pass Filter. (a) Circuit; (b) gain versus frequency characteristics.

Example

If $R_1 = 2$ kΩ, $R_2 = 4$ kΩ, $C_1 = 0.05$ μF, then

$$f_c = \frac{1}{2\pi(4 \cdot 10^3 \ \Omega)(5 \times 10^{-8} \ F)} = 796 \ Hz$$

Ideally, the circuit will amplify all voltages whose frequency is less than 796 Hz with a gain magnitude

$$|A| = \frac{R_2}{R_1} = \frac{4 \ k\Omega}{2 \ k\Omega} = 2 \cong 6 \ dB$$

The gain, above 796 Hz, decreases at a rate of 20 dB per decade of frequency.

Amplifier circuits, even simple one-amplifier ones, go on and on and on. Many publications are now available whose sole subject is amplifier applications. Figure 7-13 shows the circuit configurations and input-output equations for several additional fundamental one-amplifier circuits. They should be thoroughly studied and researched to handle the more complex circuits.

7.5. ADVANCED AMPLIFIER APPLICATIONS

Operational amplifier circuits are used to:

1. Provide *amplification.*
2. Perform *mathematical operations.*
3. *Generate* voltage signals.
4. *Convert* an input variable, i.e., charge, current, voltage, time, etc., to a voltage.
5. *Filter* AC voltages.

Function Generator

The examples given in the previous section were circuits that used only one amplifier. Most amplifier applictions use two or more. Figure 7-14 is a schematic drawing of a circuit that *generates* two voltage waveforms. It is called a *function generator*, because one of its voltage outputs (A_2) is triangular in shape and the other output, A_1, is a voltage squarewave. The generator consists of an integrator A_2 and a threshold detector or comparator A_1. The integrator implements the advanced mathematical function of integration. For a fixed positive DC voltage on its input, the output of the integrator will be a negative-going voltage ramp. The model for the integrator is a constant-current source charging a capacitor. The ramp will be positive-going for a fixed negative DC voltage on its input. The complete circuit has three feedback paths: (1) C_1 around A_2, (2) R_3 around A_1, and (3) R_2 from A_2 to A_1. C_1 and R_2 provide negative feedback, and R_3 provides positive feedback. The net feedback for the circuit is positive and is a requirement

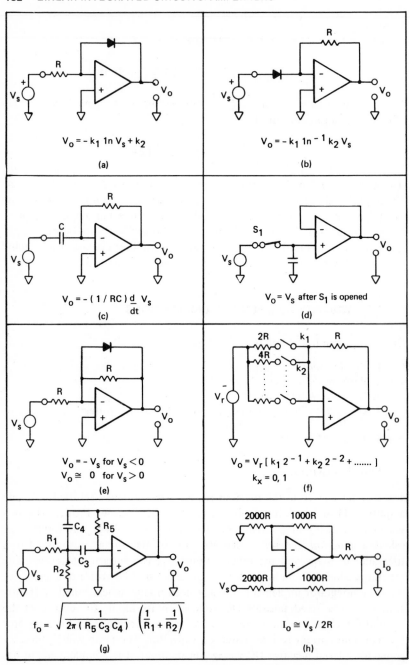

Figure 7-13. One-Amplifier Circuits. (a) Log amplifier; (b) antilog amplifier; (c) differentiator; (d) sample-and-hold; (e) half-wave rectifier; (f) digital-to-analog converter; (g) bandpass filter; (h) current mirror circuit.

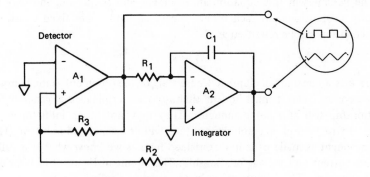

Figure 7-14. Function Generator

for regenerative circuits or those that produce a repetitive output with no input required.

To understand the operation of the circuit, let the output of A_1 be at its positive saturation level. This fixed positive DC voltage is the input of the integrator. The integrator's output will be a negative-going voltage ramp. R_2 and R_3 form a voltage divider whose end points are the positive saturation voltage of A_1 and the steadily decreasing negative-going voltage of A_2. The junction of these two resistors is sensed by the noninverting input of A_1 with the inverting input at ground or 0 V. When the noninverting input senses a slightly negative voltage (originally it was positive), the output of A_1 will swing from its positive saturation voltage to its negative saturation voltage. The voltage ramp of A_2 will now stop going negatively and will begin to go positively. It will continue to increase until the junction of R_2 and R_3 goes to slightly positive voltage with respect to ground. At this time A_1's output is switched back to the positive saturation level. This procedure is repeated continuously.

The tri-wave frequency is determined by R_1, C_1, and the positive and negative saturation voltages of the amplifier A_1. The frequency of the triangular and square voltage waveforms will be the same. The amplitude of the output voltage waveforms is established by the R_2/R_3 ratio and A_1's saturation voltages. The generator may be made independent of the saturation voltages (and hence the supply voltages) by clamping the output of A_1 with matched, back-to-back zener diodes.

Simulated Inductor

The reactance of an inductor for the AC steady-state condition is given as

$$\frac{V(t)}{I(t)} = X_L = 2\pi f L$$

The variables in this equation are reactance and frequency, and they are directly proportional to each other, since 2, π, and L are fixed constants. This equation can be rewritten as

$$X_L = kf$$

where k is a constant or number whose unit of measurement is the henry or ohm-seconds. Then it must be true that *any* two terminals that generate this relationship will also be an inductor. They may not be an inductor in the sense of wire wound on a bobbin, but they will be a simulated inductor. What a component is made of is immaterial, so long as we know what it is called and its current and voltage relationship. The circuit in Figure 7-15 is a simulated inductor. This simulated inductor circuit is called a *gyrator*, and its function is to rotate or gyrate a capacitor into an inductor. This circuit can be monolithically fabricated, and it provides the IC designer the use of inductors in his circuit designs. The analysis of the circuit is not mathematically complex, but it is tedious. The step-by-step procedure is as follows:

STEP 1: $V_{A1} = 2V_1$ (Noninverting amplifier with a gain of 2)

STEP 2: $V_{A2} = -2V_1 + 2V_2$ (Use superposition for the V_2 and V_{A1} circuit inputs)

STEP 3: $I_A = \dfrac{V_1 - V_{A1}}{R} = \dfrac{V_1 - 2V_1}{R} = \dfrac{-V_1}{R}$

$I_B = \dfrac{V_1 - V_2}{R}$

$I_C = \dfrac{V_{A2} - V_2}{R} = \dfrac{2V_2 - 2V_1 - V_2}{R} = \dfrac{V_2 - 2V_1}{R}$

STEP 4: $I_2 = I_B + I_C = \dfrac{V_1 - V_2}{R} + \dfrac{V_2 - 2V_1}{R} = \dfrac{-V_1}{R}$

STEP 5: $I_1 = I_A + I_B = -\dfrac{V_1}{R} + \dfrac{V_1 - V_2}{R} = -\dfrac{V_2}{R}$

The ratio of $\dfrac{V_1}{I_1}$ defines the impedance, or in the AC steady state the reactance, of the input terminals.

From steps 4 and 5,

$$X_{in} = \frac{V_1}{I_1} \quad \frac{-RI_2}{-V_2/R} = \frac{R^2}{V_2/I_2}$$

However, the ratio of $\dfrac{V_2}{I_2}$ defines the impedance, or in AC steady state the reactance, of the output terminals. Hence,

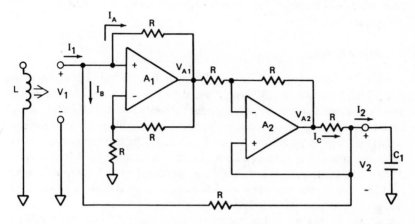

Figure 7-15. Simulated Inductor

$$X_{in} = \frac{R^2}{X_{out}}$$

$$X_L = \frac{R^2}{X_C}$$

$$X_L = \frac{R^2}{\frac{1}{2\pi fC}}$$

$$X_L = 2\pi fCR^2 = 2\pi f(CR^2) = 2\pi fL_{eq} = kf$$

The equivalent value of inductance $L_{equivalent}$, is equal to CR^2. Its unit of measurement is ohm-seconds or henries and can be verified by noting that the unit of measurement, or dimension, for RC is seconds.

$$R^2 C = R(RC) = [\text{ohm-seconds}] = [\text{henries}]$$

All discrete filter circuits that use a grounded inductor can now be implemented with monolithic techniques. The restriction of having one inductor terminal grounded is a disadvantage; however, other circuits are available that allow the simulated inductor to "float." This amplifier circuit, in its simplest sense, simulates the two-terminal passive device called an inductor.

Example

If $C = 0.1 \ \mu F$ and $R = 10 \ k\Omega$, then

$$L_{eq} = CR^2 = (10^{-7} \ F)(10^4 \ \Omega)^2$$

$$L_{eq} = 10 \ \text{Henries}$$

High values of inductance are possible if modest values of R and C are used in the gyrator circuit.

High-Pass Filter

The RL circuit of Figure 7-16(a) is a simple discrete component, *high-pass filter*. The filter will ideally pass all signal frequencies above the corner frequency and attenuate or amplitude-reduce those below it. This circuit can be intuitively analyzed by considering the reactance of L at very low frequencies and at very high frequencies. At low frequencies, the inductance reactance is small and approaches zero as a limit. X_L and R form an AC voltage divider. At low frequencies, X_L is low, and its proportional AC voltage drop is small. At high frequencies, the inductive reactance is large and ideally approaches infinity or an open. The inductor's proportional voltage drop is large and approaches V_s as a limit. The output voltage will be 70.7 percent of its input at the cutoff or corner frequency f_c.

$$f_c = \frac{2\pi R}{L}$$

The Bode plot of Figure 7-16(b) displays the high-pass filter characteristic. The RL circuit of Figure 7-17 exhibits the *same* characteristic. In this circuit, the inductance L is simulated by using the gyrator circuit. Most important, however, is the fact that the active RL filter can be integrated, which results in small size and weight and low cost.

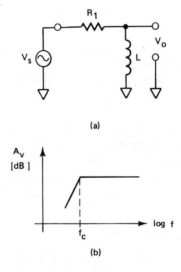

(a)

(b)

Figure 7-16. High-Pass Filter. (a) Discrete circuit; (b) gain versus frequency.

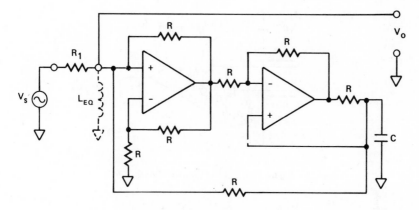

Figure 7-17. High Pass Active Filter

CURRENT-MODE OPERATIONAL AMPLIFIER

7.6. CURRENT-MODE AMPLIFIER FUNDAMENTALS

The *conventional* operational amplifier is a voltage-mode device. The input and output variables are voltage, and they are related by the amplifier's gain, *A*. Another class of monolithic amplifiers exists called *current-mode operational amplifiers*. The output variable is voltage, but the input variable is current. In a conventional amplifier, the input differential amplifier senses the difference of two voltages. In the current-mode amplifier, the input current-mirror circuit senses the difference of two currents. The name "Norton Amplifier" is used to indicate this new type of operation. It was specifically designed to operate from a *single* power supply voltage, a requirement of many industrial control systems, including automobiles.

The uniqueness of this amplifier is in the form of its input circuit, which is called a *current mirror*. This circuit, Figure 7-18, operates in the current mode, since input currents are compared or differenced. It can be thought of as a Norton differential amplifier. The voltages at the input terminals are fixed at one diode drop above ground. The current-mode amplifier is used in circuits similar to the op amp, where external components are connected to provide feedback, or the components interconnect the amplifier to signal sources and DC voltages. Signal voltages are converted to currents by using input resistors.

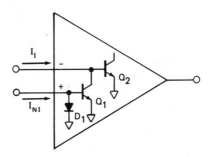

Figure 7-18. Current Mirror Circuit

The input circuit or current mirror senses current. A current I_{NI} will flow into the noninverting (+) terminal of the current mirror circuit of Figure 7-18. This current will turn D_1 on, and the (+) terminal will go to approximately +0.6 V. This voltage is impressed across the base to the emitter of Q_1 and will cause a collector current to flow of the same value as I_{NI}. This occurs because the *p-n* junctions associated with the diode and *B-E* of the transistor are monolithic. They are made at the same time, in the same die area, and under the same conditions; hence, they will exhibit the same voltage-versus-current characteristic. The collector current of Q_1 can be sourced only from the inverting (-) terminal. The amplifier output, through the feedback components, forces the condition of I_I equals I_{NI}. The base currents of Q_1 and Q_2 are much smaller than the I_{NI} and I_I currents and represent error currents. However, the true signal current of the amplifier is the base current of Q_2, which is relatively small. With Q_2 on, the voltage at the inverting (-) input will equal its base emitter voltage and is labeled V_I. The amplifier output and input are related through the voltage gain from V_I to V_0.

A new symbol is used for the current-mode amplifier. The current arrow, Figure 7-19, on the (+) input is used to indicate that this functions as the current input. The current source symbol between the inputs implies the current mode of operation and that current is removed from the (-) input. Included in this figure is the first-order circuit model. In the model, the two diodes fix the voltage of the inputs to approximately 0.6 V. The current I_{NI}, determined by external sources and resistors, is mirrored by the current source connected to the (-) input. Its value is established by the current in the noninverting input. I_B represents the base current of Q_2. The amplifier output voltage equals the voltage gain A_V times the inverting input voltage V_I and is symbolized by a dependent voltage source.

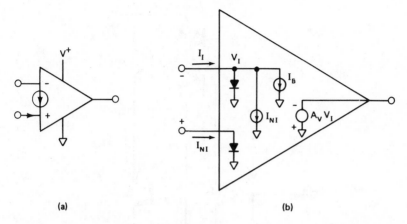

Figure 7-19. Current Mode Amplifier. (a) Schematic symbol; (b) circuit model.

7.7. INTEGRATED CURRENT-MODE AMPLIFIER

The schematic for a single amplifier of the quad 3900 is shown in Figure 7-20. Because of the circuit simplicity, four of these amplifiers are usually fabricated on a single chip. One common biasing circuit is used for all of the individual amplifiers. All of the voltage gain is provided by the common emitter circuit of Q_2. This stage achieves a large voltage gain (70 dB) through use of an active load implemented by Q_5, which is a current source, and the gain of Q_4. Q_4 also functions as an interstage buffer to reduce the loading at the high impedance collector of Q_2. Transistor Q_7 is an output emitter follower and serves as a driver for the load currents. The collector-base junction of Q_4 becomes forward-biased under a large negative output voltage swing condition. This transistor converts to a vertical *pnp* during this mode of operation, which causes the output to change from class A bias to class B. This allows the amplifier to sink more current than that provided by the Q_6 current source. The transistor Q_3 provides the class B action, which exists under large signal operating conditions.

The complete schematic, Figure 7-21, of the current-mode 3900 amplifier includes the four amplifiers, input clamps, turn-on circuitry, and current source biasing. Q_1, Q_2, Q_4, and Q_5 are the *pnp*, 200 μA, current sources. These multiple-collector devices have one collector tied to the base to fix its current gain. The base current for these transistors is predetermined through Q_{28}. The base and emitter voltages of this transistor are fixed; hence the emitter and collector currents are established. One-fourth of this current

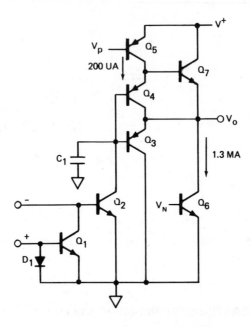

Figure 7-20. Schematic of a Simplified 3900 Amplifier

is distributed to each *pnp* current source base. The emitter current of Q_{28} voltage biases, through Q_{33}, the 1.3 mA *npn* current sources of Q_{31}, Q_{32}, Q_{34}, and Q_{35}. Components Q_{20}, Q_{30}, and D_6 "start up" the biasing circuits. The input clamps are provided by the multiple emitter transistor Q_{21}. These clamps protect the amplifier inputs against damage from negative voltages.

7.8. CURRENT-MODE AMPLIFIER APPLICATIONS

AC Amplifier: Inverting

The Norton amplifier readily lends itself to use as an *AC amplifier*, because the output can be biased to any desired DC level within the range of the output voltage swing. The (+) input terminal DC current, Figure 7-22,

$$I_{NI} = \frac{V^+ - 0.6 \text{ V}}{R_3}$$

This current is mirrored at the (–) terminal and must be sourced by the amplifier output. The coupling capacitor blocks the DC signal current. The output DC voltage

$$V_0 \text{ (DC)} = + 0.6 \text{ V} + I_{NI} \cdot R_2 = 0.6 \text{ V} + \frac{V^+ - 0.6 \text{ V}}{R_3} R_2$$

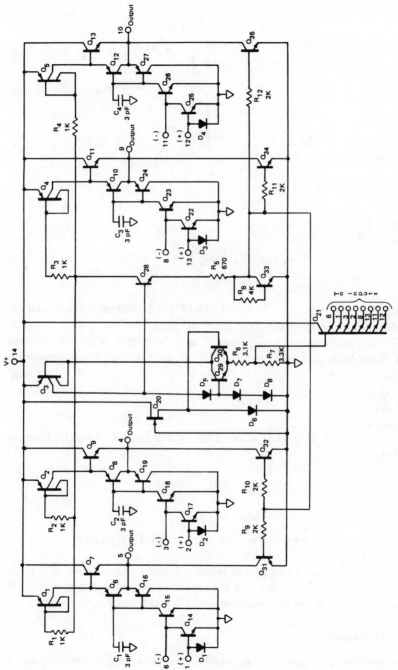

Figure 7-21. Schematic of the 3900 IC Current-Mode Amplifier.

141

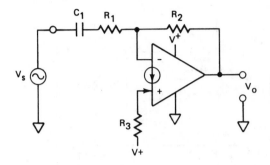

Figure 7-22. AC Amplifier: Inverting

If $2R_2 = R_3$,

$$V_0 \text{ (DC)} = 0.6 \text{ V} + V^+/2 - 0.3 \text{ V}$$

$$\cong V^+/2, \text{ since } V^+ \gg 0.3 \text{ V}$$

Thus, the amplifier output is biased to a DC voltage between ground and the positive supply and will allow maximum AC signal transfer.

The (−) terminal is the equivalent of an AC common, or ground, and thus the closed-loop gain is determined in the same manner as an operational amplifier.

$$A_{VCL} \text{ (AC)} = -R_2/R_1$$

Example

If $V^+ = +12 \text{ V}, R_1 = 3 \text{ k}\Omega, R_2 = 12 \text{ k}\Omega, R_3 = 24 \text{ k}\Omega$, then $V_0 \text{ (DC)} \cong +6 \text{ V}$ and the gain

$$A_{CL} \text{ (AC)} = -\frac{12 \text{ k}\Omega}{3 \text{ k}\Omega}$$

$$A_{CL} \text{ (AC)} = -4$$

For a 1-V peak AC input voltage, the output voltage will swing 4 V peak from +2 V to +10 V.

The negative sign in the gain equation indicates a phase reversal at the output. The value of C_1 is such that the capacitive reactance is extremely small compared to R_1 in the frequency range of the input signal.

AC Amplifier: Noninverting

The amplifier in Figure 7-23 shows both a *noninverting AC amplifier* and a second method for DC biasing. By making $R_2 = R_3$, V_0 (DC) will be equal

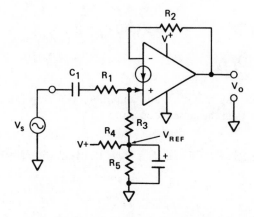

Figure 7-23. AC Amplifier: Non-Inverting

to the reference voltage that is applied to the resistor R_3. By making $R_4 = R_5$ and much smaller in value than R_2 and R_3, the reference voltage will be equal to $V^+ / 2$. The filtered $V^+ / 2$ reference shown can also be used for other amplifiers (remember, Norton amplifiers come four to a package). The (-) terminal current

$$I_I = I_{NI} \cong \frac{V_{REF} - 0.6 \text{ V}}{R_3}$$

and

$$V_0 \text{ (DC)} = 0.6 \text{ V} + I_{NI} \cdot R_2 = 0.6 \text{ V} + \frac{(V_{REF} - 0.6 \text{ V})}{R_3} R_2$$

For $R_2 = R_3$

$$V_0 \text{ (DC)} = V_{REF} = V^+ / 2$$

V_{REF} is established by the lower-valued resistor divider of R_4 and R_5, and can be any value.

The AC current through R_1 is summed with the I_{NI} DC current and reflected to the output through R_2. Hence

$$A_{CL} \text{ (AC)} \cong R_2 / R_1$$

The AC resistance of the amplifier input diode is a potential source of error and can be large for small currents. The following AC gain equation accounts for this diode resistance r_d.

$$A_V(AC) = \frac{R_2}{R_1 + r_d}$$

where $r_d = \dfrac{0.026}{I_{NI}}$ Ω

Of course, for any amplifier, the amplifier open-loop gain must be significantly greater than the closed-loop gain.

Example

If $V^+ = +12$ V, $R_1 = 6$ kΩ, $R_2 = R_3 = 12$ kΩ, $R_4 = R_5 = 1$ kΩ, then

$$V_0 \text{ (DC)} = V^+ / 2 = +6V$$

and

$$A_{CL} \text{ (AC)} \cong R_2 / R_1 = 2$$

The current $I_{NI} \cong 6$ V / 12 kΩ = 0.5 mA and the diode resistance,

$$r_d = 0.026 \text{ V} / 0.0005 \text{ A} = 52 \ \Omega,$$

which is small compared to R_1 but represents an error of approximately 1 percent.

Comparator

A *comparator* with hysteresis is called a Schmitt trigger. This type of comparator has different trip points for its high and low outputs, thus displaying a hysteresis effect. The inverting and noninverting configurations using Norton amplifiers are shown with their corresponding transfer (V_0 versus V_{in}) characteristics in Figure 7-24. In each case, positive feedback is provided by R_3.

In the inverting circuit, the comparator output voltage is high as V_{in} increases from 0 to +9.5 V. For this range of input voltage, the (+) input current

$$I_{NI} = I_2 + I_3$$

$$= \frac{V^+ - V_{BE}}{R_2} + \frac{V_{SAT} - V_{BE}}{R_3}$$

For the component and source values given, I_{NI} is approximately 9 μA and the (–) input current

$$I_I = \frac{V_{in} - V_{BE}}{R_1}$$

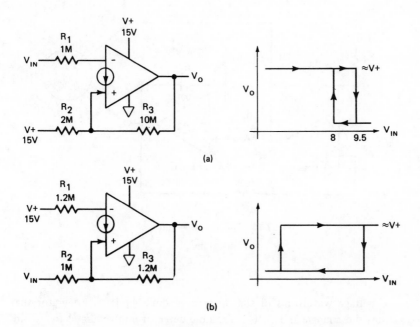

Figure 7-24. Norton Comparators with Hysteresis. (a) inverting; (b) noninverting.

will be less than I_{NI} for V_{in} less than 9.5 V. At 9.5 V I_I is greater than I_{NI} and V_0 switches to the low state.

The comparator voltage is low as V_{in} is decreased from some high positive value to +8 V. For this range of input voltage, the (+) input current

$$I_{NI} = I_2 + I_3$$

$$= \frac{V^+ - V_{BE}}{R_2} + 0$$

$$= 7.5 \ \mu A$$

The (−) input current will be greater than (+) input current for V_{in} greater than 8 V. At 8 V, I_I is less than I_{NI} and V_0 switches to the high state.

The noninverting circuit operates in a similar manner.

Function Generator

Most standard op amp circuits can be converted to an equivalent Norton amplifier configuration. The schematic in Figure 7-25 is a tri-wave or *function generator*. The outputs of this circuit are the same as the triangular and

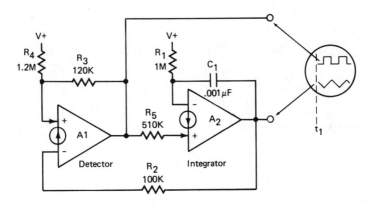

Figure 7-25. Function Generator

square voltage waveforms of the op amp version. A_2 is an integrator that generates the triangle and A_1 is a threshold detector that generates the square wave.

Although the two amplifier versions produce similar voltage waveforms and use a detector and integrator, their detailed circuit operation is different. This is due to fundamental differences in each amplifier's mode of operation. In the op amp circuit, the two integrator slopes were achieved by reversing the direction of the current by using the positive and negative saturation voltages of the detector. In the current-mode amplifier circuit, the slopes are achieved by using the current mirror feature.

The output of the detector at t_1 (Figure 7-25) is low. The capacitor C_1 is charged by the current of R_1, and its output goes in the negative direction. When the integrator output voltage falls to a point where the inverting input current of the detector is less than the noninverting current, the detector output switches from the low- to high-voltage state. The high voltage at the detector output and R_5 forces the integrator's capacitor current to change direction because of the current mirror. A_2's voltage then increases until the state of A_1 is again forced to switch. The process continually repeats itself.

If $R_1 = 2R_5$, the waveform will have good symmetry. For this condition

$$f_0 = \frac{V^+ - V_{BE}}{2R_1 C_1 \Delta V_0}$$

where ΔV_0 is the difference between the trip points of the detector. This detector is a Schmitt-trigger type.

QUESTIONS

1. What are the three main characteristics of an operational amplifier? Compare their values with those of discrete components or circuits.
2. What are the basic stages of an operational amplifier? What are the functions of each stage?
3. Does every linear circuit need an input? List one example.
4. Why is feedback employed in amplifier circuits? Name the two types of feedback and how they are achieved.
5. If we idealize an operational amplifier, we establish two amplifier input constraints. What are they?
6. What is the minimum number of pins that an operational amplifier must have? What is the typical number found in its IC form? Define them.
7. Identify the differences between the schematic symbols of the operational and current-mode amplifiers. Identify the similarities.
8. Identify the function of the voltage differential amplifier stage and the current mirror circuit.
9. What was the specific purpose for designing the current-mode IC amplifier? List several applications.
10. Identify the circuit similarities and differences of the op amp and Norton amplifier AC inverting amplifiers.

PROBLEMS

1. If the open-loop gain of an amplifier is unity at 500 kHz, what is the open-loop gain at 5 kHz? Assume that the amplifier's frequency response is due to one capacitor. What reasonable maximum closed-loop gain can we achieve at 5 kHz?
2. Specify reasonable values for R_1 and R_2 for a gain of 10 in the noninverting amplifier circuit of Figure 7-7. What is the maximum current that the amplifier must deliver if there is no load resistor and the limit of the amplifier's linear range is 10 V?
3. Specify reasonable values for R_1, R_2, R_3, and R_4 in the summing amplifier circuit of Figure 7-8 to implement the following input/output relationship.

$$V_0 = - [V_{s1} + 2V_{s2} + 3V_{s3} + 4V_{s4}]$$

4. Connect two standard op amp circuits together to form a difference configuration. Specify the circuit conditions for an input/output relationship of

$$V_0 = V_{s2} - V_{s1}$$

5. Sketch the graph of the following functions.
   (a) $V_0 = kt$
   (b) $V_0 = kt^2$

6. What is the corner frequency for the filter of Figure 7-12 if $R_2 = 10 \text{ k}\Omega$ and $C_1 = 0.01 \text{ } \mu\text{F}$? What is the value of R_1 if a gain of 10 is desired at low frequencies?

7. Specify a set of circuit values in Figure 7-15 for a simulated inductance of 0.1 H. Comment on a discrete inductor of this value.

8. What are the R_1 and R_2 values in Figure 7-22 if $R_3 = 5 \text{ k}\Omega$, $V^+ = +12 \text{ V}$, A_{CL} (AC) = 10, and the output is biased at +6 V DC? Comment on the required value of C_1.

9. Calculate the trip points of the noninverting Norton comparator in Figure 7-24(b).

10. What should the ratio of R_3 to R_4 and R_5 be in Figure 7-23 to minimize the loading of V_{REF}? List an example.

8

Linear Integrated Circuits
Regulators and Comparators

Regulators are linear integrated circuits that establish and maintain a constant DC output voltage independent of input voltage and output current variations.

Comparators are specialized amplifiers that compare two voltage inputs, decide which is greater, and then display the decision with one of two logic levels at its output.

This chapter discusses the linear integrated circuits called regulator and comparator.

REGULATORS

8.1. INTRODUCTION

The design and choice of circuits used in electronic equipment vary greatly, depending on the equipment's ultimate application or usage. Primarily, digital circuits are used in computers, RF circuits in communications equipment, and analog circuits in linear control systems. However, as diverse as the equipment designs are, they have a common requirement. They must all have a minimum of one power source. A small number of the systems may have their circuitry powered by batteries or an AC source, but the greatest requirement is for well-regulated DC voltage supplies. The supplies must provide a constant voltage to the electronic circuits and hold its value independent of changes in the input line AC, load current, or even temperature. The portion of the power supply that maintains this constant voltage is called a *voltage regulator*.

The initial appearance of monolithic voltage regulators was noteworthy. Their usage saved space and was highly cost effective. However, the earlier versions still required numerous components, in addition to the IC, to complete the design of the regulator circuit. Today monolithic voltage regulator ICs come complete. They have an input that accepts a nonregulated voltage and an output that provides a well-regulated DC voltage. The third pin of the three-pin package is the ground or reference for the input and output voltage. Most three-terminal regulators have fixed DC output voltages but can be made adjustable by using external resistors. After the operational amplifier, it is the most popular linear IC.

8.2. POWER SUPPLY

Figure 8-1 is a block diagram of a general case power supply. Power supplies are necessary to convert the AC voltage of our electric utility to the specific DC voltage required in electronics equipment. In converting AC to DC, the peak magnitude of the AC has to be only moderately greater than the DC output. The transformer steps down, or occasionally steps up, the AC line voltage to accommodate the requirement. Probably, a more important function of the transformer in power supplies is to provide isolation between the utility AC and the equipment. This is necessary to prevent the utility from excessive loading under fault conditions. The AC is then rectified or changed to pulsating DC by a full-wave or half-wave rectifier. The voltage is smoothed by a passive π or T filter. At this point, a load could be connected. In fact, for simple applications with minimal load requirements, many power supply designs stop here. However, any piece of electronics equipment that requires only a fraction of an ampere or a clean DC voltage will require a voltage regulator.

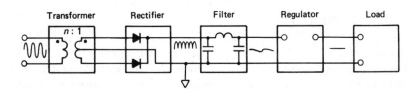

Figure 8-1. General Case Power Supply

8.3. REGULATOR FUNDAMENTALS

Regulator Function

The primary function of any regulator is to maintain a constant, predetermined voltage over an expected range of load current. It must do this for changing input voltage, output current, and temperature.

Characteristics

The primary characteristics of a regulator are constant output voltage, wide range of output current, and good line and load regulation. Secondary characteristics include input voltage range, ripple rejection, power dissipation, quiescent current, noise, and stability. The regulator is a variable-voltage input, constant-voltage output device. It will maintain this relationship for various changing conditions within the operating specifications of the device. *Load regulation* refers to the change in output voltage for a change in load current. *Line regulation* refers to the change in output voltage for a change in input voltage. Both characteristics are expressed as percentages and reflect the ability of a regulator to maintain a constant voltage output with a change in input voltage or output current. Table 8-1 lists the numbers associated with the primary characteristics of three, three-terminal, monolithic positive regulators. Cost and performance are related, and the application and its economics determines the proper selection.

TABLE 8-1

	Regulator		
Characteristic	309K	340 - 15	78L24
Output voltage	5V ± .2 V	15 V ± . 6 V	24 V ± 1 V
Input voltage	7 – 30 V	17.5 – 30 V	27 – 38 V
Output current	0 – 1.5 A	0 – 1.5 A	0 – .1 A
Load regulation	1 percent max	0.2 percent max	0.83 percent max
Line regulation	1 percent max	0.2 percent max	1.5 percent max

The regulator portion of the supply can be implemented by a three-terminal regulator such as that shown in Figure 8-2. Regulators are housed in various packages to accommodate different load current requirements. The power and thermal capabilities of the package and the pass element size primarily determine the output current capability of the regulator. The T05 can nominally deliver 200 mA maximum, and the T03 can deliver 1 A. All

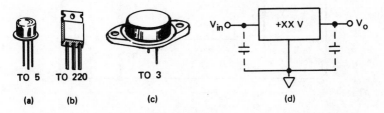

TO 5   TO 220   TO 3

(a)   (b)   (c)   (d)

Figure 8-2. The Three-Terminal Positive IC Regulator. (a) - (c) Packages; (d) symbol.

packages depend on adequate heat sinking. The regulator is symbolically represented by a rectangle with function, value, or identifying number written inside. The input and output of the regulator are usually bypassed by small capacitors for frequency stability. This is required if the regulator is remote from the power supply filter. The low cost and small size of regulators make the monolithic regulator well suited for local regulation. In most electronic systems the power supply, up to the filter, is located in one assembly and parallel regulators, one for each printed circuit board, are located in the other assemblies. Local regulation provides for better load regulation because of the shorter leads, smaller load currents, and reduced line capacitance and inductance.

8.4. INTEGRATED POSITIVE REGULATOR

The simplified diagram for a positive voltage regulator is shown in Figure 8-3. This series-type regulator is a basic proven design and lends itself well to integration. The regulator consists of a pass transistor Q_1 and error amplifier A_1, and a reference voltage V_Z. The reference establishes a value for the regulator's output voltage. The output voltage will be proportional to the reference voltage, with R_1 and R_2 establishing the constant of proportionality.

$$V_o = V_Z \left(\frac{R_1 + R_2}{R_1} \right) = k V_Z$$

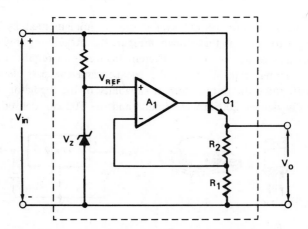

Figure 8-3. Simplified Positive Voltage Regulator.

Error Amplifier and Pass Transistor

For three-terminal regulators, resistors R_1 and R_2 are part of the device. The error amplifier is a simplified voltage amplifier. It compares a portion of the output voltage, divided by R_1 and R_2, with the reference voltage. When an error or difference exists, it drives the base of Q_1 to compensate for the difference. The emitter of Q_1, which is the output, follows the base. If the output voltage drops below the established value (kV_Z) because of load conditions, A_1's output will go more positive, raising the output voltage. If the output voltage rises above the established value, A_1 will go less positive, lowering the output voltage.

Q_1 is the regulator pass transistor. This transistor passes the required output current. Its collector voltage is the nonregulated DC input, and its emitter voltage is the regulated DC output. The transistor's V_{CE} makes up the difference between the two and can be large. The load current is supplied by V_{in} through Q_1 and along with V_{CE} determines the power dissipated by the pass element. Q_1 is designed as a power transistor and occupies over half of the IC chip in the higher-current regulators.

Voltage Reference

The performance of the voltage reference is crucial to the overall performance of the regulator. It is the basis for comparison by the error amplifier in establishing and regulating the output voltage. The noise, stability, and thermal effects of the reference are directly reflected to the regulator's output. The simplest voltage reference is a zener diode. Monolithic voltage regulators use either a zener diode or a complex circuit that can be *modeled* as a zener diode. Both can produce a precise and stable reference voltage.

Protection Features

The proper functioning of the power supply is vital to the operation of electronics equipment and systems. If it is "down," the entire system is inoperable, since all other circuits require the supply's voltage and current to operate. Troubleshooting any part of the system cannot begin until the power supplies are "up." It is also a function that is subjected to the adverse conditions of high temperature, high power dissipation, and excessive loading. Included in the regulator design are features to safeguard the device against these conditions. Figure 8-4 is a simplified regulator circuit and includes the main self-protection features. The three features illustrated are

1. Thermal shutdown (Q_4).
2. Excessive power dissipation protection $(D_1, R_3,$ and $R_4)$.
3. Current limiting $(Q_3$ and $R_{CL})$.

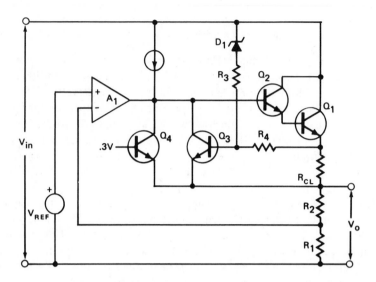

Figure 8-4. Simplified Positive Voltage Regulator with Self-Protection Features.

The basic regulator circuit has been modified to more closely represent the actual IC design. The output of the error amplifier drives the pass transistor, but its pullup is from a current source, not previously shown. V_{REF} is now shown as an independent source. Components Q_3, Q_4, D_1, R_4, and R_{CL} have been added to show the protective features of the regulator. Most monolithic regulators today include in their design current limiting, excessive power dissipation, and thermal shutdown.

Current limiting is a method of limiting output current to avoid regulator failure resulting from large output current loads or dead shorts. The current-limiting transistor, Q_3, is normally off. Under a heavy load or short condition, the voltage drop across R_{CL} increases enough to turn on Q_3, which, in turn, sinks the base drive current that is being supplied to Q_1. The amount of base drive is limited. With less Q_1 base current, Q_1 cannot conduct as much collector current and is limited until the adverse output condition has been corrected. This technique is similar to that used in the output stage of operational amplifiers.

Excessive power dissipation protection is intended to protect the output pass transistor from breakdown caused by excessive V_{CE} at high currents. If the input voltage, V_{in}, becomes too great, D_1 breaks down. This will cause current to be drawn through R_4, lowering the amount of current it takes through R_{CL} to turn on the current-limiting transistor Q_3. The voltage drops of R_4 and R_{CL} *add*. The pass transistor has been converted to a Darlington configuration. The addition of Q_2 decreases the loading of the error amplifier.

Thermal shutdown prevents the IC chip from overheating in case of excessive load, momentary short, or increase in ambient temperature. When the chip reaches a temperature of 165°C to 180°C, the output current is limited to prevent further heating. This protective feature uses the breakdown voltage temperature coefficient of the forward-biased junction of Q_4 (about -2.0 mV/°C). The base of Q_4 is biased to about 0.3 V, too low to turn it on at room temperature. As the die temperature exceeds the prespecified value, the turn-on voltage of Q_4 approaches 0.3 V, and it begins to conduct. This again shunts the base drive to Q_1 and limits output current.

Detailed

The complete schematic for a monolithic 340 is shown in Figure 8-5. In design and performance, it is similar to two other popular positive regulators, the 340L and 78LXX.

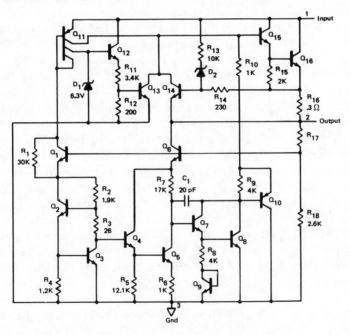

Figure 8-5. Schematic of the 340 IC Regulator.

This device contains 16 transistors, 2 diodes, and 18 resistors. Included in the above components are *npns, pnps,* multiple-collector lateral *pnps,* a capacitor and diffused resistors. The functions and their associated components follow.

1. Pass transistor (Darlington): Q_{16}, Q_{15}, R_{15}.
2. Error amplifier: Q_{11}, R_{10}, R_9, Q_{10}, Q_8, Q_7, R_8, Q_9, C_1, Q_6.
3. Voltage reference: Q_{11}, R_1, Q_1, R_2, Q_2, R_3, Q_3, R_4, Q_4, R_5, Q_5, R_6, R_7.
4. Current limiting: Q_{14}, R_{16}.
5. Excessive power dissipation: D_2, R_{13}, R_{14}.
6. Thermal shutdown: D_1, Q_{12}, R_{11}, R_{12}, Q_{13}.

The voltage gain for the regulating loop, symbolized in the simplified diagram by an amplifier symbol, is provided by the common emitter stage of Q_8. It is configured as a Darlington pair with Q_7, which buffers its input. The collector load for Q_8 is a current source whose value is established by the base-to-emitter voltage of Q_{10} across R_9. This current is sourced from the multiple-collector *pnp*, Q_{11}, through R_{10}. The voltage at the junction of R_{10} and the collector of Q_{11} drives the pass transistors Q_{15} and Q_{16}. The capacitor, C_1, is connected between the collector and base of the Q_7-Q_8 gain stage. It provides negative feedback and frequency stability for the circuit.

Resistors R_{17} and R_{18} are the gain-setting resistors used in establishing the value of output voltage. They are the equivalent of R_2 and R_1 in the regulator simplified diagram. The junction of the two resistors is connected to the base of Q_6, whose emitter voltage is the temperature-stable V_{REF}.

The collectors of Q_{13} and Q_{14} are connected to the base of the pass transistor pair. Q_{13} and Q_{14} are normally off and will turn on and shunt the base current of the pass transistors under fault conditions. R_{16} is the current-limiting resistor (R_{CL}). When the voltage drop across this resistor approaches 0.6 V, Q_{14} will turn on and limit the output current of the regulator.

Components D_2, R_{13}, and R_{14} implement the excessive power dissipation protection circuit. When the input voltage to the regulator becomes excessive, D_2 will break down and cause current to flow through R_{13} and R_{14}. The voltage drop across R_{14} will cause less of a voltage drop across R_{16} necessary to turn the current-limiting transistor Q_{14} on.

Transistor Q_{13} will turn on when the chip temperature exceeds a predetermined value, approximately $175°C$. The reference voltage for the base of this transistor is developed with a conventional zener diode, D_1. This voltage is buffered by Q_{12} and voltage-divided with R_{11} and R_{10}. The divided voltage, about 300 mV, is the base bias voltage of Q_{13}. The emitter base voltage of Q_{13} is the actual temperature sensor. As temperature increases, the turn-on base-to-emitter voltage of the transistor decreases at a rate of about -2 mV/°C. As V_{BE} approaches the base bias voltage, it will turn on and shunt the pass transistor's base current.

Q_1, Q_2, Q_3, Q_4, Q_5 and associated components establish the reference voltage at the emitter of Q_6. The reference voltage must be temperature-stable and in actuality is the sum of two voltages with opposing temperature coefficients. This circuit technique of generating a voltage reference is called a "band gap" or ΔV_{BE} reference. In the simplified diagram, it is modeled as a zener diode. The circuit operation is somewhat complex and is redrawn in simplified form in Figure 8-6. In this circuit Q_1 is connected as a diode whose forward voltage is V_{BE} and whose current is high compared to Q_2. Because of the different emitter currents of Q_1 and Q_2, the emitter voltage of Q_2 will be equal to their differences $(V_{BE2} - V_{BE1})$ or ΔV_{BE}. The voltage drop across R_2 will be R_2/R_2 times ΔV_{BE}. The reference voltage at the collector of Q_3 will be the sum of $V_{BE}(Q_3)$ plus $(R_2/R_2) \Delta V_{BE}$. If the gain R_2/R_2 is properly chosen, the negative temperature coefficient (TC), of $V_{BE}(Q_3)$ can be made to cancel the positive TC of ΔV_{BE}, producing nearly zero temperature drift for the reference.

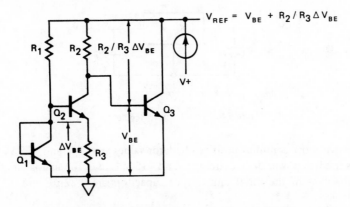

Figure 8-6. Simplified Schematic of the Band Gap Reference.

Three-terminal regulators also use zener diode references. These diodes are subject to adverse electrical effects within the chip and are placed below the die surface with a new technology known as *ion implantation*. Figure 8-7 shows the integrated structure of a reference zener diode. Band gap references are generally chosen for higher-current devices (0.5 A - 3 A), where they offer low noise without significantly increasing the die area, while zeners are chosen for small-die, lower-current (0.1 A - 0.25 A) devices.

Q_{11} is a multiple-collector *pnp* transistor.

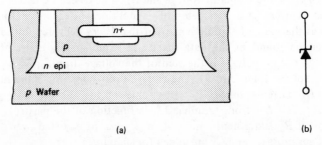

(a)   (b)

Figure 8-7. Regulator Zener Reference. (a) Integrated structure; (b) schematic symbol.

8.5. POSITIVE REGULATOR APPLICATIONS

Fixed Output Regulator

The monolithic, three-terminal regulator is complete and ready to use. Apply a nonregulated input voltage, and the device produces a fixed, regulated output voltage. Figure 8-8 illustrates the simplicity of its application. An input and output capacitor are added to aid frequency stability and transient response of the circuit.

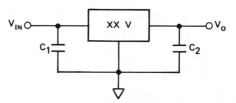

Figure 8-8. Fixed Output Regulator.

The two most common positive regulator values are +5 V and +15 V. Five volts is used to power digital circuitry, and +15 V is used for analog circuitry. Typical values of the input and output capacitors are 0.22 μF and 0.1 μF, respectively.

Adjustable Output Regulator

Although positive regulators that have various output voltage values can be purchased, many times odd voltages or variable voltages are required. The addition of two resistors, R_1 and R_2 in Figure 8-9, converts the fixed output regulator to a variable output. The output voltage of the regulator circuit

$$V_0 = V_{REG} + \left(\frac{V_{REG}}{R_1} + I_Q \right) R_2$$

where V_{REG} is the regulator voltage and I_Q is the device's quiescent current. Quiescent current is a parameter of monolithic voltage regulators and is defined as that part of the input current that is *not* delivered to the load. Its value can be found in the data sheet and is moderately constant. In this circuit, the output voltage is the sum of the voltage drop across R_2 and the regulator voltage (pins 2 to 3). The voltage drop across R_2 is the sum of the I_1 and I_Q currents times R_2. The value of the current I_1 is equal to V_{REG}/R_1, and I_Q, from terminal 3, is a function of the particular type of regulator. If R_2 is a potentiometer, then V_0 can be adjusted over a range of values. Capacitors C_1 and C_2 are added for stability.

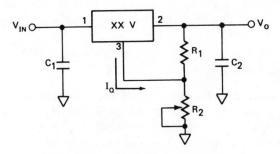

Figure 8-9. Adjustable Output Regulator.

Example. The typical quiescent current for an 8-V 340 regulator is 8 mA. If R_1 is a fixed 1-kΩ resistor and R_2 is a 250-Ω potentiometer, then

$$V_0 = 8 \text{ V} + \left(\frac{8 \text{ V}}{1 \text{ k}\Omega} + 8 \text{ mA} \right) \cdot R_2$$

and the output voltage can be adjusted from +8 V to +12 V.

Current Regulator

The regulator is a voltage-in, voltage-out device. However, it can be configured as a voltage-in, current-out device. The voltage between pins 2 and 3 of the regulator, Figure 8-10, is a fixed value. If a resistor R_1 is added, then the constant voltage across the resistor will develop a constant current. The only major error is the quiescent current of pin 3, but it is a known constant value. Hence

$$I_0 = \frac{V_{REG}}{R_1} + I_Q$$

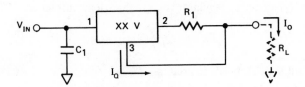

Figure 8-10. Current Regulator.

Example. For a +12-V regulator, $I_Q = 8$ mA, $R_1 = 1$ mA, the output current

$$I_0 = \frac{12 \text{ V}}{1 \text{ k}\Omega} + 8 \text{ mA} = 20 \text{ mA}.$$

This circuit will deliver a constant 20 mA to the load. The maximum load voltage will depend on the maximum unregulated input voltage to the regulator.

High-Current Voltage Regulator

Large electronic systems use power supplies that deliver extremely heavy load currents. The voltage regulator must also be capable of handling this heavy current. Monolithic voltage regulators are usually not capable of more than 5 amperes of current because of the power dissipation of the pass transistor and its lower gain at the higher current level. These problems are circumvented by using an external transistor, as shown in the 10-A regulator of Figure 8-11. At low current levels, Q_1 is off. The current is sourced by V_{in} through R_1 and the regulator to the load. When the current through R_1 develops a voltage drop of about 0.6 V, Q_1 will turn on and *shunt* the excess current to the load. Voltage regulation at the output is still performed by the regulator.

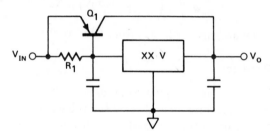

Figure 8-11. High-Current Voltage Regulator.

High-Current Regulator with Short-Circuit Protection

The circuit in Figure 8-12 provides a high-output current and takes advantage of the internal current limiting of the regulator to provide short-circuit current protection for the booster transistor as well. If the diode voltage V_{D1} equals the V_{BE} of Q_1, then the regulator and Q_1 share load current according to

$$I_1 = \frac{R_2}{R_1} I_{REG}$$

Similarly, during output short circuits

$$I_1 \text{ (SC)} = \frac{R_2}{R_1} I_{REG} \text{ (SC)}$$

If the thermal characteristics of the regulator and Q_1 are similar, the thermal protection of the regulator will be extended to Q_1. Resistor R_3 decreases the response time of Q_1 to load changes.

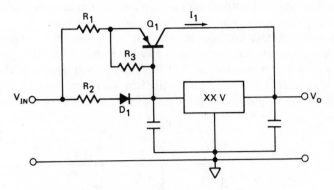

Figure 8-12. High-Current Regulator with Short-Circuit Protection.

8.6. NEGATIVE REGULATOR APPLICATIONS

Fixed Output Regulator

The application of negative three-terminal regulators is similar to that of positive three-terminal regulators. Apply a negative nonregulated input voltage, and the device produces a fixed, negative, regulated output voltage. Figure 8-13 illustrates the simplicity of its application. An input and an output capacitor are added to aid frequency stability and transient response.

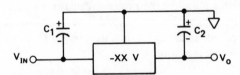

Figure 8-13. Fixed Output Regulator.

The most common negative regulator values are –5 V and –15 V. Negative five volts is used to power digital circuitry (ECL), and –15 V is used to power analog circuitry. Typical values of the input and output solid tantalum capacitors are 2.2 μF and 1 μF, respectively.

The 320 is one of several popular negative regulators. All their designs, in principle, are similar.

Adjustable Output Regulator

Although negative regulators that have various output voltage values can be purchased, many times odd voltages or variable voltages are required. The technique of using a resistor divider to increase the output voltage in the positive regulator circuit can also be used with negative regulators.

The output voltage may also be raised by simply placing a zener diode in series with the ground pin, as indicated in Figure 8-14. The ground or quiescent current, 1 mA for the 320, biases the zener diode on. The zener diode must have a low-temperature coefficient where low output voltage drift is required. The output voltage is the sum of the zener and regulator voltage

$$V_0 = V_R + V_Z$$

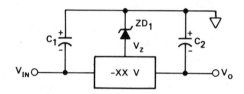

Figure 8-14. Fixed Output Regulator of Odd Value.

Basic Dual Supplies

Figure 8-15 is a schematic diagram of a simple dual output supply with the output voltage equal to the preset value of the regulators. Diodes D_2 and D_3 protect the supply against polarity reversal of the outputs during overloads. Diode D_1 and resistor R_1 allow the positive regulator to start up when $+V_{in}$ is delayed relative to $-V_{in}$ with a heavy load drawn between the outputs. Most positive regulators will latch up during start-up, with a heavy load current flowing to the negative regulator.

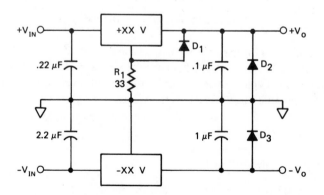

Figure 8-15. Fixed Dual Supply.

The addition of three resistors and two trim potentiometers (Figure 8-16) converts the fixed dual supply to one whose outputs can be adjusted to a tighter tolerance.

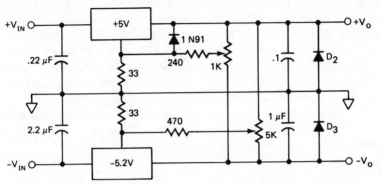

Figure 8-16. Trimmable Dual Supply.

Variable Output Lab Supply

The circuit in Figure 8-17 will provide 0 to –20 V output with up to 1 A load current. It uses the 120 as a pass element, a 301A as an external error amplifier and a 113 as the voltage reference. The regulator is short-circuit-proof, and, together with the amplifier, features superior regulation characteristics at all voltages.

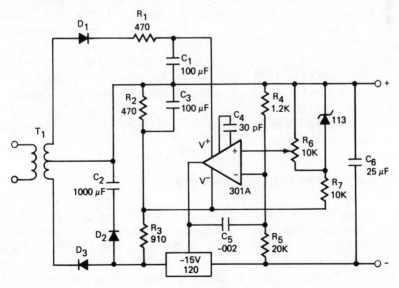

Figure 8-17. Lab Supply, 0- -20 V.

Transformer T_1 steps down the line AC voltage and has a tapped secondary to provide the positive voltage for the op amp. D_1 and D_3 are half-wave rectifiers. The positive rectified output of D_1 is filtered by R_1 and C_1 and applied to the V^+ pin of the op amp. The negative rectified output of D_3 is filtered and connected to the nonregulated input of the regulator. The ground pin of the regulator is driven by the error amplifier, whose voltage reference at the noninverting input is established by R_6 and the 113. The error amplifier will voltage-drive the regulator until the voltage at the amplifier inverting input is approximately the same as the voltage established at the noninverting input. The voltage at the inverting input is the output voltage divided by R_4 and R_5. Capacitors C_4 and C_5 provide frequency compensation and stability for the amplifier. The output filter capacitor C_6 prevents overshoot after momentary shorts and minimizes ripple and noise.

COMPARATORS

8.7. INTRODUCTION

Most electronic systems contain analog *and* digital circuits. Both types do not exist as separate entities but are eventually joined together by specialized circuits called *interface circuits*. These circuits are not purely digital or linear but contain both functions. They convert one form to the other. Modern electronic systems are computer- or microprocessor-controlled. Coded digital information is transmitted to and from the computer to a portion of the system where the test, measurement, detection, and control occur. This portion is usually implemented with linear devices. The system-level interface devices that directly convert digital information to analog information and analog to digital are called D/A and A/D converters. These are usually modular or hybrid, but a few are monolithic. At the circuit level, a comparator is used to convert analog to digital.

8.8. COMPARATOR FUNDAMENTALS

The comparator is a two-input, one-output voltage-amplifying linear device that is capable of high gain, high input resistance, and low output resistance. This definition is the same for an operational amplifier. In fact, the operational amplifier can be used as a comparator. However, a comparator performs a different function. It detects two input voltages and provides an output that has two discrete states. The two states indicate whether one input is greater in magnitude than the other, or vice versa. The comparator differs from the operational amplifier in four ways:

1. It performs a different function.
2. It is operated open-loop.
3. It is significantly faster.
4. It has two output states whose voltage levels are digitally compatible.

The operational amplifier operates closed-loop. The two inputs track each other, and the output is a voltage that is linearly related to the input signal. The output can take on any value within its operational (saturation voltages) limits. The comparator operates open-loop. It has differential amplifier inputs, but normally one input is connected to a voltage reference. The other input is connected to a voltage that varies with time. The two comparator inputs do not track each other. In fact, the two are the same only at the time the comparator changes its output state. The output states are based on whether one input is greater or less than the other input. Two output voltage levels represent the two states of the decision. Since the saturation voltage levels of op amps are not compatible with digital circuits, the output stage of most comparators is specifically designed to provide +5 V and 0 V. The values between these two voltages carry no information and occur only during the brief transition from one state to the other.

The response time for op amps is in the tens of microseconds. This is extremely slow compared to the operating time of digital circuits and systems. The comparator is designed for minimum response time, and although it is not quite as fast as a digital device, it is many orders of magnitude faster than operational amplifiers.

The schematic symbol for a comparator, Figure 8-18, is the same as for the amplifier. Both have two inputs and two power supply pins. However, most comparators also have ground and strobe pins. The ground pin is used to bias the output stage for digital logic compatibility. The strobe pin is used to enable the comparator to make a decision based on its inputs or disable the

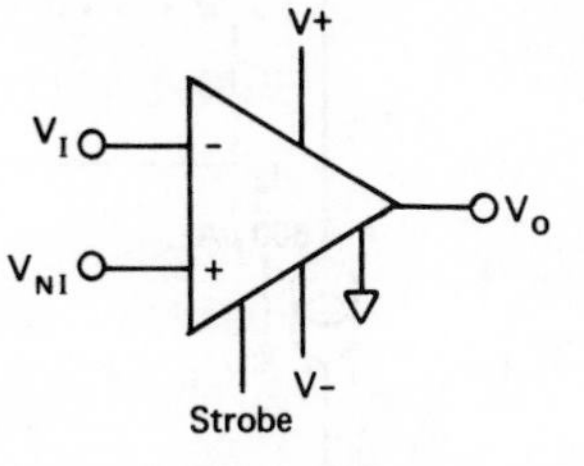

$$V_I > V_{NI} \quad V_O = \text{LOW}$$
$$V_I < V_{NI} \quad V_O = \text{HIGH}$$

(a)

(b)

Figure 8-18. IC Comparator. (a) Schematic symbol; (b) mathematical model.

comparator and have its output go to a predetermined permanent state. Two more pins complete the typical eight-pin package, and they are used to null out (balance) the inherent comparator input error. Specifically, this is called the *offset voltage* and it represents the mismatch of the V_{BE}s of the differential amplifier stage transistors.

8.9. INTEGRATED COMPARATOR

The simplified schematic of the 311 comparator is shown in Figure 8-19. The input differential amplifier is implemented with the Q_1-Q_3 and Q_2-Q_4 transistor pairs. This stage differences two voltages, provides gain, and minimizes the input error current. The output of this stage is further amplified by the

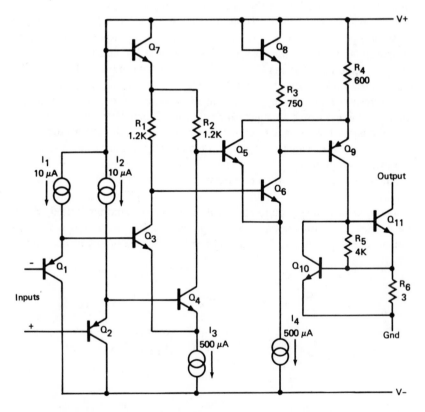

Figure 8-19. Simplified Schematic of the 311 IC Comparator.

Q_5-Q_6 pair. This stage feeds Q_9, which provides additional gain and drives the output stage. The circuit current sources are used to determine the biasing so that performance is not greatly affected by supply voltage changes. The output transistor is Q_{11}, and it is protected by Q_{10} and Q_9, which limit the peak output current. Q_{10}, normally off, will turn on when the voltage drop across R_6 approaches 0.6 V. With Q_{10} on, the base current of Q_{11} is shunted to the output, where current limiting takes place. The output lead, since it is not connected to any other point, can be returned to a positive supply through a pull-up resistor. This supply is usually +5 V but is not limited to this value. The output stage can also be biased to interface with circuits that operate from 0 to a negative voltage. This can be accomplished by grounding the output pin through a pull-up resistor, and then returning the ground pin to a negative voltage.

The complete schematic of the comparator is shown in Figure 8-20.

8.10. COMPARATOR APPLICATIONS

Detector with Strobe

The voltage sources, V_{s1} and V_{s2}, in Figure 8-21(a) can represent time-varying signal sources or DC voltages. If one is used as a basis for comparison, it is called a *reference voltage*. When the comparator is *enabled* through its strobe, the output

$$V_0 = +5 \text{ V} \quad \text{for } V_{s1} < V_{s2}$$

and

$$V_0 = 0 \text{ V} \quad \text{for } V_{s1} > V_{s2}$$

Resistor R_1 is the pull-up resistor for the open-collector transistor in the output stage of the comparator. Q_1 and R_2 are external components necessary to interface a digital logic signal to the device's strobe pin. If the comparator is disabled or inhibited via the strobe signal, the output will go to a pre-determined permanent state (the output transistor is off for the 311).

When one of the input sources is at zero volts or an input pin is ground, the circuit is referred to as a *zero crossing detector*. The output will change state each time the signal source crosses zero volts to the opposite voltage polarity.

The transfer characteristics of Figure 8-21(b) illustrate the use of this circuit as a reference detector. The source V_{s2} is a fixed DC reference source. The comparator output will be low for any value of V_{s1} greater than V_{s2} and high for the opposite condition.

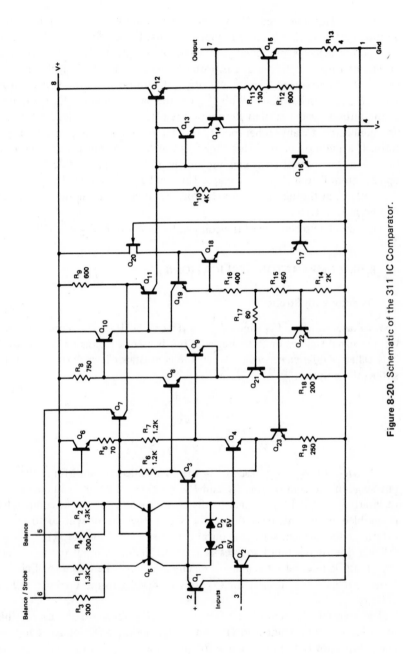

Figure 8-20. Schematic of the 311 IC Comparator.

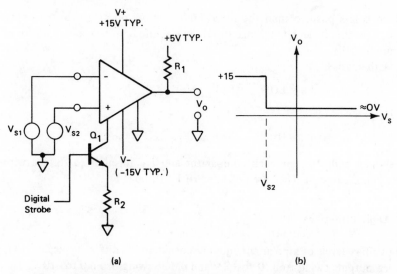

Figure 8-21. Detector with Strobe. (a) Circuit; (b) transfer characteristics.

Window Comparator

The use of two devices, connected as shown in Figure 8-2(a), produces a window comparator. This circuit provides a low output voltage if the signal source V_s is more positive than the lower limit (LL) voltage and less positive than the upper limit (UL) voltage. The circuit output is high if the signal

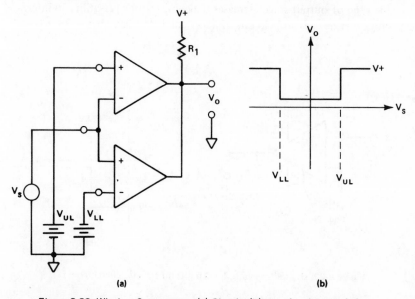

Figure 8-22. Window Comparator. (a) Circuit; (b) transfer characteristics.

source is less positive than the lower limit or more positive than the upper limit. The circuit detects a range or window of input voltage and is shown in Figure 8-22(b).

Mathematically,

$$V_0 = \text{LOW} \quad \text{for } V_{LL} < V_S < V_{UL}$$

and

$$V_0 = \text{HIGH} \quad \text{for } V_s > V_{UL} \text{ or } V_s > V_{LL}$$

The open collectors of each comparator are tied together to form a wire ORed configuration. They are pulled up to V^+ through R_1.

Digital Interface

The voltage levels of bipolar and unipolar logic families differ. Typically, TTL device outputs swing from 0 to +5 V and pMOS swings from 0 to -10 V. The circuit of Figure 8-23 joins or interfaces the two logic families. V_s models the output of a TTL device. When V_s is low or near 0 V, the more positive voltage (+2.5 V) on the inverting input will cause the comparator output to go to its low state, which is near -10 V. The comparator ground pin, which essentially is the emitter of the output stage transistor, is connected to -10 V. When V_s is high or near +5 V, the more positive voltage or the noninverting input will cause the comparator output to go to its high state, 0 V, through the pull-up resistor R_1.

This type of output stage increases the comparator's versatility in interfacing with other devices and circuits, notably solid state switches.

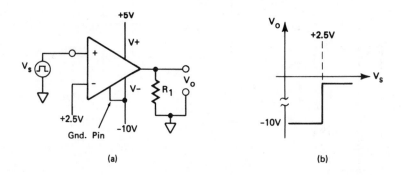

(a) (b)

Figure 8-23. Digital Interface. (a) Circuit; (b) transfer characteristics.

Free-Running Multivibrator

Comparators, occasionally, are operated closed-loop. The circuit of Figure 8-24 is a squarewave oscillator with positive (R_4) and negative (R_3) feedback. The comparator does not operate in its linear region, but switches from one saturation voltage level to the other. The time at which the output switches is a function of the two input voltages. The noninverting (+) input voltage is DC and is one of two values depending on the comparator output voltage. The inverting (–) input is a time-varying voltage due to the charging and discharging of the capacitor C_1 through R_3. If the output is at the high level, V^+,

$$V_{NI} = \frac{R_2}{R_2 + R_1//R_4} \cdot V^+$$

and if the output is at the low level, $\simeq 0$ V,

$$V_{NI} = \frac{R_2//R_4}{R_2//R_4 + R_1} \cdot V^+$$

For V^+ equal to +5 V and the component values given,

$$V_{NI} \cong +3 \text{ V}, +2 \text{ V}$$

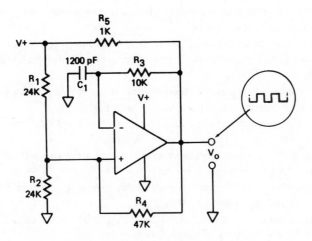

Figure 8-24. Free-running Multivibrator.

The capacitor is either being charged through R_3 and the comparator output to V^+, or is being discharged because the comparator output is at zero volts.

When the output is at +5 V (through R_5), the noninverting (+) input is at +3 V. The inverting (−) input is charging toward +3 V. When it reaches +3 V, the comparator output switches to near zero volts and the noninverting input voltage goes to +2 V. The capacitor then discharges toward 0 V, but when it reaches +2 V, the output switches back to +5 V and the process repeats itself.

Components C_1 and R_3 determine the frequency of oscillation. The time period of the oscillating frequency is approximately

$$T = \frac{1.1}{\tau} , \quad \text{where } \tau = R_3 C_1$$

The ratio of the resistors R_1 and R_2, with the proper R_4, determines the symmetry or duty cycle of the waveform. Equal values produce a symmetric waveform with the high and low voltage times equal. Resistor R_5 is the pull-up resistor and must be small compared to R_3 and R_4.

QUESTIONS

1. What is the function of each of the five blocks in the general case power supply of Figure 8-1?
2. List the primary characteristics of a voltage regulator. List the secondary characteristics of the voltage regulator.
3. List the advantages provided by a monolithic regulator if it locally regulates a DC voltage on a printed circuit board.
4. Define line and load regulation. Identify the variables and the units of measurement for each case.
5. Why is the performance of the voltage reference in a regulator crucial to the regulator's performance? How many types of references are generally used?
6. How many self-protection features does a regulator typically have? What does each feature protect against?
7. Why is it necessary to use a Darlington configured pass element in the positive voltage regulator shown in Figure 8-4?
8. What are the most common positive and negative voltage regulator values? List an application for each case.
9. Modify the block diagram of the positive regulator in Figure 8-3 to make it a negative regulator.
10. What important regulator characteristic will be improved if the gain of A_1 in Figure 8-3 is significantly increased?
11. Why are input and output capacitors required in monolithic regulator circuits?

12. What is the typical maximum load current for a monolithic regulator?
13. List the functional differences between an operational amplifier and a comparator. List the similarities.
14. Why do comparators have a ground pin? Is this terminal restricted to ground?
15. What function does the comparator strobe pin perform?

PROBLEMS

1. In the simplified positive regulator circuit of Figure 8-3, determine the output voltage if V_Z = 1.5 V, R_1 = 2 kΩ, and R_2 = 18 kΩ. What is the smallest load resistor that the regulator can drive if its rated output is 200 mA maximum?

2. The thermal shutdown transistor Q_4 shown in Figure 8-4 is biased to 0.350 V at 25°C, its B-E barrier potential is 0.600 V, and the B-E temperature coefficient is -2.2 mV/°C. At what temperature will the device turn on?

3. The current-limiting resistor R_{CL} in Figure 8-4 is 0.30 Ω. The barrier potential of Q_3 is 0.660 V. If Q_3's base current is neglected, at what value of load current will Q_3 turn on and begin to cause the regulator to current-limit?

4. Estimate a value of the reference voltage in Figure 8-6 if R_2 = 6 kΩ and R_3 = 600 Ω.

5. The typical slewing rate of an op amp is 0.5 V/μsec and for a comparator is 100 V/μsec. If both devices swing 10 V, how much faster is the comparator?

Appendix A

Data sheets of representative ICs used in this book

DIGITAL INTEGRATED CIRCUITS

Device Number	Manufacturer	Technology	Function
SN6400/5400	Texas Instruments	bipolar-TTL	NAND gate
MC10101	Motorola	bipolar-ECL	NOR/OR gate
CD4001	RCA	unipolar-CMOS	NOR gate
8080A	Intel	unipolar-NMOS	Microprocessor
2102A	Intel	unipolar-NMOS	RAM

LINEAR INTEGRATED CIRCUITS

Device Number	Manufacturer	Function
uA741C	Fairchild	operational amplifier
LM301A	National	operational amplifier
LM311	National	comparator
LM3900	National	current-mode amplifier
LM340	National	voltage regulator

CIRCUIT TYPES SN5400, SN7400
QUADRUPLE 2-INPUT POSITIVE NAND GATES

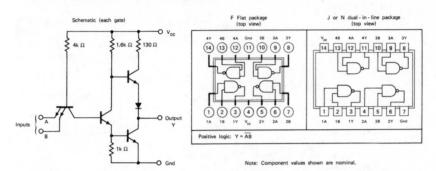

Positive logic: Y = $\overline{AB}$

Note: Component values shown are nominal.

recommended operating conditions

	MIN	NOM	MAX	UNIT
Supply voltage V_{CC}: SN5400 Circuits	4.5	5	5.5	V
SN7400 Circuits	4.75	5	5.25	V
Normalized Fan-Out From Each Output, N.			10	
Operating Free-Air Temperature Range, T_A: SN5400 Circuits	−55	25	125	°C
SN7400 Circuits	0	25	70	°C

electrical characteristics (over recommended operating free-air temperature range unless otherwise noted)

	PARAMETER	TEST FIG.	TEST CONDITIONS[†]		MIN	TYP[‡]	MAX	UNIT
$V_{in}(1)$	Logical 1 input voltage required at both input terminals to ensure logical 0 level at output	1	V_{CC} = MIN		2			V
$V_{in}(0)$	Logical 0 input voltage required at either input terminal to ensure logical 1 level at output	2	V_{CC} = MIN				0.8	V
$V_{out}(1)$	Logical 1 output voltage	2	V_{CC} = MIN, V_{in} = 0.8 V, I_{load} = −400μA		2.4	3.3		V
$V_{out}(0)$	Logical 0 output voltage	1	V_{CC} = MIn, V_{in} = 2 V, I_{sink} = 16 mA			0.22	0.4	V
$I_{in}(0)$	Logical 0 level input current (each input)	3	V_{CC} = MAX, V_{in} = 0.4 V				−1.6	mA
$I_{in}(1)$	Logical 1 level input current (each input)	4	V_{CC} = MAX, V_{in} = 2.4 V				40	μA
			V_{CC} = MAX, V_{in} = 5.5 V				1	mA
I_{OS}	Short-circuit output current §	5	V_{CC} = MAX	SN5400	−20		−55	mA
				SN7400	−18		−55	
$I_{CC}(0)$	Logical 0 level supply current	6	V_{CC} = MAX, V_{in} = 5V			12	22	mA
$I_{CC}(1)$	Logical 1 level supply current	6	V_{CC} = MAX, V_{in} = 0			4	8	mA

switching characteristics, V_{CC} = 5 V, T_A = 25° C, N = 10

	PARAMETER	TEST FIG.	TEST CONDITIONS	MIN	TYP	MAX	UNIT
t_{pd0}	Propagation delay time to logical 0 level	65	C_L = 15 pF, R_L = 400Ω		7	15	ns
t_{pd1}	Propagation delay time to logical 1 level	65	C_L = 15pF, R_L = 400Ω		11	22	ns

[†] For conditions shown as MIN or MAX, use the appropriate value specified under recommended operating conditions for the applicable device type.

[‡] All typical values are at V_{CC} = 5V, T_A = 25° C

§ Not more than one output should be shorted at a time.

QUAD OR/NOR GATE   **MECL 10,000 SERIES MOTOROLA**

MC10101

The MC10101 is a quad 2-input OR/NOR gate with one input from each gate common to pin 12. Input pulldown resistors eliminate the need to tie unused inputs to an external supply.

POSITIVE LOGIC   NEGATIVE LOGIC

P_D = 25 mW typ/gate (No Load)
t_{pd} = 2.0 ns typ
Output Rise and Fall Time:
= 3.5 ns typ (10% - 90%)
= 2.0 ns typ (20% - 80%)

V_{CC1} = Pin 1
V_{CC2} = Pin 16
V_{EE} = Pin 8

CIRCUIT SCHEMATIC

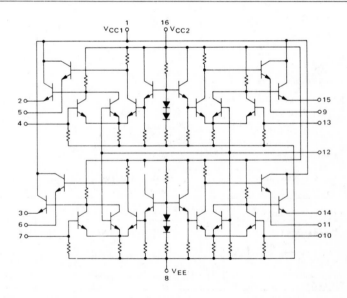

ELECTRICAL CHARACTERISTICS

Each MECL 10,000 series has been designed to meet the dc specifications shown in the test table, after thermal equilibrium has been established. The circuit is in a test socket or mounted on a printed circuit board and transverse air flow greater than 500 linear fpm is maintained. Outputs are terminated through a 50-ohm resistor to −2.0 volts. Test procedures are shown for only one gate. The other gates are tested in the same manner.

L SUFFIX
CERAMIC PACKAGE
CASE 620

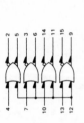

TEST VOLTAGE VALUES
(Volts)

	$V_{IH max}$	$V_{IL min}$	$V_{IHA min}$	$V_{ILA max}$	V_{EE}	(V_{CC}) Gnd
@ Test Temperature −30°C	−0.890	−1.890	−1.205	−1.500	−5.2	1,16
+25°C	−0.810	−1.850	−1.105	−1.475	−5.2	1,16
+85°C	−0.700	−1.825	−1.035	−1.440	−5.2	1,16

MC10101L Test Limits

Characteristic	Symbol	Pin Under Test	−30°C Min	−30°C Max	+25°C Min	+25°C Typ	+25°C Max	+85°C Min	+85°C Max	Unit	TEST VOLTAGE APPLIED TO PINS LISTED BELOW: $V_{IH max}$	$V_{IL min}$	$V_{IHA min}$	$V_{ILA max}$	V_{EE}	(V_{CC}) Gnd
Power Supply Drain Current	I_E	8	—	—	—	20	26	—	—	mAdc					8	1,16
Input Current	I_{inH}	4	—	—	—	—	265	—	—	μAdc	4				8	1,16
		12	—	—	—	—	535	—	—	μAdc	12				8	1,16
	I_{inL}	4	—	—	—	—	—	—	—	μAdc		4			8	1,16
		12	—	—	—	—	—	—	—	μAdc		12			8	1,16
Logic "1" Output Voltage	V_{OH}	5	−1.060	−0.890	0.5	−0.960	−0.810	−0.890	−0.700	Vdc	12				8	1,16
		5	−1.060	−0.890	0.5	−0.960	−0.810	−0.890	−0.700	Vdc	4					
		2	−1.060	−0.890		−0.960	−0.810	−0.890	−0.700	Vdc	—					
		2	−1.060	−0.890		−0.960	−0.810	−0.890	−0.700	Vdc	—					
Logic "0" Output Voltage	V_{OL}	5	−1.890	−1.675		−1.850	−1.650	−1.825	−1.615	Vdc	12				8	1,16
		5	−1.890	−1.675		−1.850	−1.650	−1.825	−1.615	Vdc	4					
		2	−1.890	−1.675		−1.850	−1.650	−1.825	−1.615	Vdc	12					
		2	−1.890	−1.675		−1.850	−1.650	−1.825	−1.615	Vdc	4					
Logic "1" Threshold Voltage	V_{OHA}	5	−1.080			−0.980		−0.910		Vdc			12		8	1,16
		5	−1.080			−0.980		−0.910					4			
		2	−1.080			−0.980		−0.910						12		
		2	−1.080			−0.980		−0.910						4		
Logic "0" Threshold Voltage	V_{OLA}	5		−1.655			−1.630		−1.595	Vdc			12		8	1,16
		5		−1.655			−1.630		−1.595					4			
		2		−1.655			−1.630		−1.595					12		
		2		−1.655			−1.630		−1.595					4		

Switching Times (50-ohm load)

											Pulse In	Pulse Out	V_{EE}	(V_{CC}) Gnd
Propagation Delay	$t_{4+2−}$	2	1.0	3.1	1.0	2.0	2.9	1.0	3.3	ns	4	2	−3.2 V	+2.0 V
	$t_{4−2−}$	2										2	8	1,16
	t_{4+5+}	5										5		
	$t_{4−5}$	5										5		
Rise Time (20 to 80%)	t_{2+}	2	1.1	3.6	1.1		3.3	1.1	3.7			2		
	t_{5+}	5										5		
Fall Time (20 to 80%)	$t_{2−}$	2										2		
	$t_{5−}$	5										5		

Pins: 4, 2, 5, 3, 6, 7, 14, 10, 11, 13, 15, 12, 9

CD4001   quadruple 2-input NOR gate

general description

The CD4001M/CD4001C is a monolithic complementary MOS (CMOS) quadruple two-input NOR gate integrated circuit. N and P-channel enhancement mode transistors provide a symmetrical circuit with output swings essentially equal to the supply voltage. This results in high noise immunity over a wide supply voltage range. No dc power other than that caused by leakage current is consumed during static conditions.

All inputs are protected against static discharge and latching conditions.

features

- Wide supply voltage range   3V to 15V
- Low power   10 nW (typ)
- High noise immunity   0.45 V_{DD} (typ)

schematic and connection diagrams

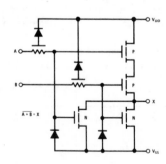

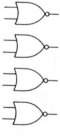

CD4001
Quad 2-Input
NOR Gate

absolute maximum ratings

Voltage at Any Pin (Note 1)	$V_{SS} - 0.3V$ to $V_{DD} + 0.3V$
Operating Temperature Range	
CD4001M	$-55°C$ to $+125°C$
CD4001C	$-40°C$ to $+85°C$
Storage Temperature Range	$-65°C$ to $+150°C$
Package Dissipation	500 mW
Operating V_{DD} Range	$V_{SS} + 3.0V$ to $V_{SS} + 15V$
Lead Temperature (Soldering, 10 seconds)	$300°C$

dc electrical characteristics CD4001M

PARAMETER	CONDITIONS	LIMITS									UNITS
		-55°C			25°C			125°C			
		MIN	TYP	MAX	MIN	TYP	MAX	MIN	TYP	MAX	
Quiescent Device	$V_{DD} = 5V$			0.05		0.001	0.05			3	µA
Current (I_L)	$V_{DD} = 10V$			0.1		0.001	0.1			6	µA
Quiescent Device Dissi-	$V_{DD} = 5V$			0.25		0.005	0.25			15	µW
pation/Package (P_D)	$V_{DD} = 10V$			1		0.01	1			60	µW
Output Voltage Low	$V_{DD} = 5V,\ V_I = V_{DD}, I_O = 0A$			0.01		0	0.01			0.05	V
Level (V_{OL})	$V_{DD} = 10V, V_I = V_{DD}, I_O = 0A$			0.01		0	0.01			0.05	V
Output Voltage High	$V_{DD} = 5V,\ V_I = V_{SS}, I_O = 0A$	4.99			4.99	5		4.95			V
Level (V_{OH})	$V_{DD} = 10V, V_I = V_{SS}, I_O = 0A$	9.99			9.99	10		9.95			V
Noise Immunity	$V_{DD} = 5V,\ V_O = 3.6V, I_O = 0A$	1.5			1.5	2.25		1.4			V
(V_{NL}) (All Inputs)	$V_{DD} = 10V, V_O = 7.2V, I_O = 0A$	3			3	4.5		2.9			V
Noise Immunity	$V_{DD} = 5V,\ V_O = 0.95V, I_O = 0A$	1.4			1.5	2.25		1.5			V
(V_{NH}) (All Inputs)	$V_{DD} = 10V, V_O = 2.9V,\ I_O = 0A$	2.9			3	4.5		3			V
Output Drive Current	$V_{DD} = 5V,\ V_O = 0.4V, V_I = V_{DD}$	0.5			0.40	1		0.28			mA
N-Channel ($I_D N$)	$V_{DD} = 10V, V_O = 0.5V, V_I = V_{DD}$	1.1			0.9	2.5		0.65			mA
Output Drive Current	$V_{DD} = 5V,\ V_O = 2.5V, V_I = V_{SS}$	-0.62			-0.5	-2		-0.35			mA
P-Channel ($I_D P$)	$V_{DD} = 10V, V_O = 9.5V, V_I = V_{SS}$	-0.62			-0.5	-1		-0.35			mA
Input Current (I_I)						10					pA

dc electrical characteristics CD4001C

PARAMETER	CONDITIONS	LIMITS									UNITS
		-40°C			25°C			85°C			
		MIN	TYP	MAX	MIN	TYP	MAX	MIN	TYP	MAX	
Quiescent Device	$V_{DD} = 5V$			0.5		0.005	0.5			15	µA
Current (I_L)	$V_{DD} = 10V$			5		0.005	5			30	µA
Quiescent Device Dissi-	$V_{DD} = 5V$			2.5		0.025	2.5			75	µW
pation/Package (P_D)	$V_{DD} = 10V$			50		0.05	50			300	µW
Output Voltage Low	$V_{DD} = 5V,\ V_I = V_{DD}, I_O = 0A$			0.01		0	0.01			0.05	V
Level (V_{OL})	$V_{DD} = 10V, V_I = V_{DD}, I_O = 0A$			0.01		0	0.01			0.05	V
Output Voltage High	$V_{DD} = 5V,\ V_I = V_{SS}, I_O = 0A$	4.99			4.99	5		4.95			V
Level (V_{OH})	$V_{DD} = 10V, V_I = V_{SS}, I_O = 0A$	9.99			9.99	10		9.95			V
Noise Immunity	$V_{DD} = 5V,\ V_O = 3.6V, I_O = 0A$	1.5			1.5	2.25		1.4			V
(V_{NL}) (All Inputs)	$V_{DD} = 10V, V_O = 7.2V, I_O = 0A$	3			3	4.5		2.9			V
Noise Immunity	$V_{DD} = 5V,\ V_O = 0.95V, I_O = 0A$	1.4			1.5	2.25		1.5			V
(V_{NH}) (All Inputs)	$V_{DD} = 10V, V_O = 2.9V,\ I_O = 0A$	2.9			3	4.5		3			V
Output Drive Current	$V_{DD} = 5V,\ V_O = 0.4V, V_I = V_{DD}$	0.35			0.3	1		0.24			mA
N-Channel ($I_D N$)	$V_{DD} = 10V, V_O = 0.5V, V_I = V_{DD}$	0.72			0.6	2.5		0.48			mA
Output Drive Current	$V_{DD} = 5V,\ V_O = 2.5V, V_I = V_{SS}$	-0.35			-0.3	-2		-0.24			mA
P-Channel ($I_D P$)	$V_{DD} = 10V, V_O = 9.5V, V_I = V_{SS}$	-0.3			-0.25	-1		-0.2			mA
Input Current (I_I)						10					pA

Note 1: This device should not be connected to circuits with the power on because high transient voltages may cause permanent damage.

8080A
SINGLE CHIP 8-BIT N-CHANNEL MICROPROCESSOR

The 8080A is functionally and electrically compatible with the Intel® 8080.

- ■ **TTL Drive Capability**
- ■ **2 μs Instruction Cycle**
- ■ **Powerful Problem Solving Instruction Set**
- ■ **Six General Purpose Registers and an Accumulator**
- ■ **Sixteen Bit Program Counter for Directly Addressing up to 64K Bytes of Memory**

- ■ **Sixteen Bit Stack Pointer and Stack Manipulation Instructions for Rapid Switching of the Program Environment**
- ■ **Decimal,Binary and Double Precision Arithmetic**
- ■ **Ability to Provide Priority Vectored Interrupts**
- ■ **512 Directly Addressed I/O Ports**

The Intel® 8080A is a complete 8-bit parallel central processing unit (CPU). It is fabricated on a single LSI chip using Intel's n-channel silicon gate MOS process. This offers the user a high performance solution to control and processing applications.

The 8080A contains six 8-bit general purpose working registers and an accumulator. The six general purpose registers may be addressed individually or in pairs providing both single and double precision operators. Arithmetic and logical instructions set or reset four testable flags. A fifth flag provides decimal arithmetic operation.

The 8080A has an external stack feature wherein any portion of memory may be used as a last in/first out stack to store/retrieve the contents of the accumulator, flags, program counter and all of the six general purpose registers. The sixteen bit stack pointer controls the addressing of this external stack. This stack gives the 8080A the ability to easily handle multiple level priority interrupts by rapidly storing and restoring processor status. It also provides almost unlimited subroutine nesting.

This microprocessor has been designed to simplify systems design. Separate 16-line address and 8-line bi-directional data busses are used to facilitate easy interface to memory and I/O. Signals to control the interface to memory and I/O are provided directly by the 8080A. Ultimate control of the address and data busses resides with the HOLD signal. It provides the ability to suspend processor operation and force the address and data busses into a high impedance state. This permits OR-tying these busses with other controlling devices for (DMA) direct memory access or multi-processor operation.

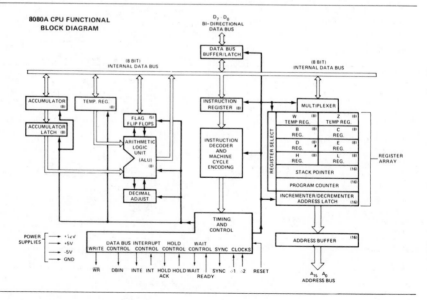

8080A CPU FUNCTIONAL BLOCK DIAGRAM

8080A

8080A FUNCTIONAL PIN DEFINITION

The following describes the function of all of the 8080A I/O pins. Several of the descriptions refer to internal timing periods.

A_{15}-A_0 (output three-state)
ADDRESS BUS; the address bus provides the address to memory (up to 64K 8-bit words) or denotes the I/O device number for up to 256 input and 256 output devices. A_0 is the least significant address bit.

D_7-D_0 (input/output three-state)
DATA BUS; the data bus provides bi-directional communication between the CPU, memory, and I/O devices for instructions and data transfers. Also, during the first clock cycle of each machine cycle, the 8080A outputs a status word on the data bus that describes the current machine cycle. D_0 is the least significant bit.

SYNC (output)
SYNCHRONIZING SIGNAL; the SYNC pin provides a signal to indicate the beginning of each machine cycle.

DBIN (output)
DATA BUS IN; the DBIN signal indicates to external circuits that the data bus is in the input mode. This signal should be used to enable the gating of data onto the 8080A data bus from memory or I/O.

READY (input)
READY; the READY signal indicates to the 8080A that valid memory or input data is available on the 8080A data bus. This signal is used to synchronize the CPU with slower memory or I/O devices. If after sending an address out the 8080A does not receive a READY input, the 8080A will enter a WAIT state for as long as the READY line is low. READY can also be used to single step the CPU.

WAIT (output)
WAIT; the WAIT signal acknowledges that the CPU is in a WAIT state.

$\overline{WR}$ (output)
WRITE; the $\overline{WR}$ signal is used for memory WRITE or I/O output control. The data on the data bus is stable while the $\overline{WR}$ signal is active low ($\overline{WR}$ = 0).

HOLD (input)
HOLD; the HOLD signal requests the CPU to enter the HOLD state. The HOLD state allows an external device to gain control of the 8080A address and data bus as soon as the 8080A has completed its use of these buses for the current machine cycle. It is recognized under the following conditions:
• the CPU is in the HALT state.
• the CPU is in the T2 or TW state and the READY signal is active.
As a result of entering the HOLD state the CPU ADDRESS BUS (A_{15}-A_0) and DATA BUS (D_7-D_0) will be in their high impedance state. The CPU acknowledges its state with the HOLD ACKNOWLEDGE (HLDA) pin.

HLDA (output)
HOLD ACKNOWLEDGE; the HLDA signal appears in response to the HOLD signal and indicates that the data and address bus

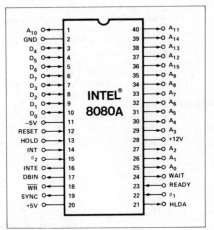

A_{10}	1	40	A_{11}
GND	2	39	A_{14}
D_4	3	38	A_{13}
D_5	4	37	A_{12}
D_6	5	36	A_{15}
D_7	6	35	A_9
D_3	7	34	A_8
D_2	8	33	A_7
D_1	9	32	A_6
D_0	10	31	A_5
−5V	11	30	A_4
RESET	12	29	A_3
HOLD	13	28	+12V
INT	14	27	A_2
ϕ_2	15	26	A_1
INTE	16	25	A_0
DBIN	17	24	WAIT
$\overline{WR}$	18	23	READY
SYNC	19	22	ϕ_1
+5V	20	21	HLDA

INTEL® 8080A

Pin Configuration

will go to the high impedance state. The HLDA signal begins at:
• T3 for READ memory or input.
• The Clock Period following T3 for WRITE memory or OUTPUT operation.
In either case, the HLDA signal appears after the rising edge of ϕ_1 and high impedance occurs after the rising edge of ϕ_2.

INTE (output)
INTERRUPT ENABLE; indicates the content of the internal interrupt enable flip/flop. This flip/flop may be set or reset by the Enable and Disable Interrupt instructions and inhibits interrupts from being accepted by the CPU when it is reset. It is automatically reset (disabling further interrupts) at time T1 of the instruction fetch cycle (M1) when an interrupt is accepted and is also reset by the RESET signal.

INT (input)
INTERRUPT REQUEST; the CPU recognizes an interrupt request on this line at the end of the current instruction or while halted. If the CPU is in the HOLD state or if the Interrupt Enable flip/flop is reset it will not honor the request.

RESET (input)[1]
RESET; while the RESET signal is activated, the content of the program counter is cleared. After RESET, the program will start at location 0 in memory. The INTE and HLDA flip/flops are also reset. Note that the flags, accumulator, stack pointer, and registers are not cleared.

V_{SS}	Ground Reference.
V_{DD}	+12 ± 5% Volts.
V_{CC}	+5 ± 5% Volts.
V_{BB}	−5 ±5% Volts (substrate bias).
ϕ_1, ϕ_2	2 externally supplied clock phases. (non TTL compatible)

8080A

INSTRUCTION SET

The accumulator group instructions include arithmetic and logical operators with direct, indirect, and immediate addressing modes.

Move, load, and store instruction groups provide the ability to move either 8 or 16 bits of data between memory, the six working registers and the accumulator using direct, indirect, and immediate addressing modes.

The ability to branch to different portions of the program is provided with jump, jump conditional, and computed jumps. Also the ability to call to and return from subroutines is provided both conditionally and unconditionally. The RESTART (or single byte call instruction) is useful for interrupt vector operation.

Double precision operators such as stack manipulation and double add instructions extend both the arithmetic and interrupt handling capability of the 8080A. The ability to

increment and decrement memory, the six general registers and the accumulator is provided as well as extended increment and decrement instructions to operate on the register pairs and stack pointer. Further capability is provided by the ability to rotate the accumulator left or right through or around the carry bit.

Input and output may be accomplished using memory addresses as I/O ports or the directly addressed I/O provided for in the 8080A instruction set.

The following special instruction group completes the 8080A instruction set: the NOP instruction, HALT to stop processor execution and the DAA instructions provide decimal arithmetic capability. STC allows the carry flag to be directly set, and the CMC instruction allows it to be complemented. CMA complements the contents of the accumulator and XCHG exchanges the contents of two 16-bit register pairs directly.

Data and Instruction Formats

Data in the 8080A is stored in the form of 8-bit binary integers. All data transfers to the system data bus will be in the same format.

D_7 D_6 D_5 D_4 D_3 D_2 D_1 D_0

DATA WORD

The program instructions may be one, two, or three bytes in length. Multiple byte instructions must be stored in successive words in program memory. The instruction formats then depend on the particular operation executed.

One Byte Instructions

D_7 D_6 D_5 D_4 D_3 D_2 D_1 D_0	OP CODE

TYPICAL INSTRUCTIONS

Register to register, memory reference, arithmetic or logical, rotate, return, push, pop, enable or disable interrupt instructions

Two Byte Instructions

D_7 D_6 D_5 D_4 D_3 D_2 D_1 D_0	OP CODE
D_7 D_6 D_5 D_4 D_3 D_2 D_1 D_0	OPERAND

Immediate mode or I/O instructions

Three Byte Instructions

D_7 D_6 D_5 D_4 D_3 D_2 D_1 D_0	OP CODE
D_7 D_6 D_5 D_4 D_3 D_2 D_1 D_0	LOW ADDRESS OR OPERAND 1
D_7 D_6 D_5 D_4 D_3 D_2 D_1 D_0	HIGH ADDRESS OR OPERAND 2

Jump, call or direct load and store instructions

For the 8080A a logic "1" is defined as a high level and a logic "0" is defined as a low level.

8080A

8080 INSTRUCTION SET

Summary of Processor Instructions

Mnemonic	Description	D_7	D_6	D_5	D_4	D_3	D_2	D_1	D_0	Clock[2] Cycles
MOVE, LOAD, AND STORE										
MOV r1,r2	Move register to register	0	1	D	D	D	S	S	S	5
MOV M,r	Move register to memory	0	1	1	1	0	S	S	S	7
MOV r,M	Move memory to register	0	1	D	D	D	1	1	0	7
MVI r	Move immediate register	0	0	D	D	D	1	1	0	7
MVI M	Move immediate memory	0	0	1	1	0	1	1	0	10
LXI B	Load immediate register Pair B & C	0	0	0	0	0	0	0	1	10
LXI D	Load immediate register Pair D & E	0	0	0	1	0	0	0	1	10
LXI H	Load immediate register Pair H & L	0	0	1	0	0	0	0	1	10
STAX B	Store A indirect	0	0	0	0	0	0	1	0	7
STAX D	Store A indirect	0	0	0	1	0	0	1	0	7
LDAX B	Load A indirect	0	0	0	0	1	0	1	0	7
LDAX D	Load A indirect	0	0	0	1	1	0	1	0	7
STA	Store A direct	0	0	1	1	0	0	1	0	13
LDA	Load A direct	0	0	1	1	1	0	1	0	13
SHLD	Store H & L direct	0	0	1	0	0	0	1	0	16
LHLD	Load H & L direct	0	0	1	0	1	0	1	0	16
XCHG	Exchange D & E, H & L Registers	1	1	1	0	1	0	1	1	4
STACK OPS										
PUSH B	Push register Pair B & C on stack	1	1	0	0	0	1	0	1	11
PUSH D	Push register Pair D & E on stack	1	1	0	1	0	1	0	1	11
PUSH H	Push register Pair H & L on stack	1	1	1	0	0	1	0	1	11
PUSH PSW	Push A and Flags on stack	1	1	1	1	0	1	0	1	11
POP B	Pop register Pair B & C off stack	1	1	0	0	0	0	0	1	10
POP D	Pop register Pair D & E off stack	1	1	0	1	0	0	0	1	10
POP H	Pop register Pair H & L off stack	1	1	1	0	0	0	0	1	10
POP PSW	Pop A and Flags off stack	1	1	1	1	0	0	0	1	10
XTHL	Exchange top of stack, H & L	1	1	1	0	0	0	1	1	18
SPHL	H & L to stack pointer	1	1	1	1	1	0	0	1	5
LXI SP	Load immediate stack pointer	0	0	1	1	0	0	0	1	10
INX SP	Increment stack pointer	0	0	1	1	0	0	1	1	5
DCX SP	Decrement stack pointer	0	0	1	1	1	0	1	1	5
JUMP										
JMP	Jump unconditional	1	1	0	0	0	0	1	1	10
JC	Jump on carry	1	1	0	1	1	0	1	0	10
JNC	Jump on no carry	1	1	0	1	0	0	1	0	10
JZ	Jump on zero	1	1	0	0	1	0	1	0	10
JNZ	Jump on no zero	1	1	0	0	0	0	1	0	10
JP	Jump on positive	1	1	1	1	0	0	1	0	10
JM	Jump on minus	1	1	1	1	1	0	1	0	10
JPE	Jump on parity even	1	1	1	0	1	0	1	0	10

Mnemonic	Description	D_7	D_6	D_5	D_4	D_3	D_2	D_1	D_0	Clock[2] Cycles
JPO	Jump on parity odd	1	1	1	0	0	0	1	0	10
PCHL	H & L to program counter	1	1	1	0	1	0	0	1	5
CALL										
CALL	Call unconditional	1	1	0	0	1	1	0	1	17
CC	Call on carry	1	1	0	1	1	1	0	0	11/17
CNC	Call on no carry	1	1	0	1	0	1	0	0	11/17
CZ	Call on zero	1	1	0	0	1	1	0	0	11/17
CNZ	Call on no zero	1	1	0	0	0	1	0	0	11/17
CP	Call on positive	1	1	1	1	0	1	0	0	11/17
CM	Call on minus	1	1	1	1	1	1	0	0	11/17
CPE	Call on parity even	1	1	1	0	1	1	0	0	11/17
CPO	Call on parity odd	1	1	1	0	0	1	0	0	11/17
RETURN										
RET	Return	1	1	0	0	1	0	0	1	10
RC	Return on carry	1	1	0	1	1	0	0	0	5/11
RNC	Return on no carry	1	1	0	1	0	0	0	0	5/11
RZ	Return on zero	1	1	0	0	1	0	0	0	5/11
RNZ	Return on no zero	1	1	0	0	0	0	0	0	5/11
RP	Return on positive	1	1	1	1	0	0	0	0	5/11
RM	Return on minus	1	1	1	1	1	0	0	0	5/11
RPE	Return on parity even	1	1	1	0	1	0	0	0	5/11
RPO	Return on parity odd	1	1	1	0	0	0	0	0	5/11
RESTART										
RST	Restart	1	1	A	A	A	1	1	1	11
INCREMENT AND DECREMENT										
INR r	Increment register	0	0	D	D	D	1	0	0	5
DCR r	Decrement register	0	0	D	D	D	1	0	1	5
INR M	Increment memory	0	0	1	1	0	1	0	0	10
DCR M	Decrement memory	0	0	1	1	0	1	0	1	10
INX B	Increment B & C registers	0	0	0	0	0	0	1	1	5
INX D	Increment D & E registers	0	0	0	1	0	0	1	1	5
INX H	Increment H & L registers	0	0	1	0	0	0	1	1	5
DCX B	Decrement B & C	0	0	0	0	1	0	1	1	5
DCX D	Decrement D & E	0	0	0	1	1	0	1	1	5
DCX H	Decrement H & L	0	0	1	0	1	0	1	1	5
ADD										
ADD r	Add register to A	1	0	0	0	0	S	S	S	4
ADC r	Add register to A with carry	1	0	0	0	1	S	S	S	4
ADD M	Add memory to A	1	0	0	0	0	1	1	0	7
ADC M	Add memory to A with carry	1	0	0	0	1	1	1	0	7
ADI	Add immediate to A	1	1	0	0	0	1	1	0	7
ACI	Add immediate to A with carry	1	1	0	0	1	1	1	0	7
DAD B	Add B & C to H & L	0	0	0	0	1	0	0	1	10
DAD D	Add D & E to H & L	0	0	0	1	1	0	0	1	10
DAD H	Add H & L to H & L	0	0	1	0	1	0	0	1	10
DAD SP	Add stack pointer to H & L	0	0	1	1	1	0	0	1	10

NOTES: 1. DDD or SSS: B 000, C 001, D 010, E 011, H 100, L 101, Memory 110, A 111.
2. Two possible cycle times, (6/12) indicate instruction cycles dependent on condition flags.

*All mnemonics copyright
© Intel Corporation 1977

8080A

8080 INSTRUCTION SET

Summary of Processor Instructions (Cont.)

Mnemonic	Description	D_7	D_6	D_5	D_4	D_3	D_2	D_1	D_0	Clock[2] Cycles
SUBTRACT										
SUB r	Subtract register from A	1	0	0	1	0	S	S	S	4
SBB r	Subtract register from A with borrow	1	0	0	1	1	S	S	S	4
SUB M	Subtract memory from A	1	0	0	1	0	1	1	0	7
SBB M	Subtract memory from A with borrow	1	0	0	1	1	1	1	0	7
SUI	Subtract immediate from A	1	1	0	1	0	1	1	0	7
SBI	Subtract immediate from A with borrow	1	1	0	1	1	1	1	0	7
LOGICAL										
ANA r	And register with A	1	0	1	0	0	S	S	S	4
XRA r	Exclusive Or register with A	1	0	1	0	1	S	S	S	4
ORA r	Or register with A	1	0	1	1	0	S	S	S	4
CMP r	Compare register with A	1	0	1	1	1	S	S	S	4
ANA M	And memory with A	1	0	1	0	0	1	1	0	7
XRA M	Exclusive Or memory with A	1	0	1	0	1	1	1	0	7
ORA M	Or memory with A	1	0	1	1	0	1	1	0	7
CMP M	Compare memory with A	1	0	1	1	1	1	1	0	7
ANI	And immediate with A	1	1	1	0	0	1	1	0	7
XRI	Exclusive Or immediate with A	1	1	1	0	1	1	1	0	7
ORI	Or immediate with A	1	1	1	1	0	1	1	0	7
CPI	Compare immediate with A	1	1	1	1	1	1	1	0	7
ROTATE										
RLC	Rotate A left	0	0	0	0	0	1	1	1	4
RRC	Rotate A right	0	0	0	0	1	1	1	1	4
RAL	Rotate A left through carry	0	0	0	1	0	1	1	1	4
RAR	Rotate A right through carry	0	0	0	1	1	1	1	1	4
SPECIALS										
CMA	Complement A	0	0	1	0	1	1	1	1	4
STC	Set carry	0	0	1	1	0	1	1	1	4
CMC	Complement carry	0	0	1	1	1	1	1	1	4
DAA	Decimal adjust A	0	0	1	0	0	1	1	1	4
INPUT/OUTPUT										
IN	Input	1	1	0	1	1	0	1	1	10
OUT	Output	1	1	0	1	0	0	1	1	10
CONTROL										
EI	Enable Interrupts	1	1	1	1	1	0	1	1	4
DI	Disable Interrupt	1	1	1	1	0	0	1	1	4
NOP	No-operation	0	0	0	0	0	0	0	0	4
HLT	Halt	0	1	1	1	0	1	1	0	7

NOTES: 1. DDD or SSS. B=000. C=001. D=010. E=011. H=100. L=101. Memory=110. A=111.
2. Two possible cycle times, (6/12) indicate instruction cycles dependent on condition flags.

intel®

2102A, 2102AL

1K x 1 BIT STATIC RAM

P/N	Standby Pwr. (mW)	Operating Pwr. (mW)	Access (ns)
2102AL-4	35	174	450
2102AL	35	174	350
2102AL-2	42	342	250
2102A-2	——	342	250
2102A	——	289	350
2102A-4	——	289	450
2102A-6	——	289	650

- **Single +5 Volts Supply Voltage**
- **Directly TTL Compatible: All Inputs and Output**
- **Standby Power Mode (2102AL)**
- **Three-State Output: OR-Tie Capability**

- **Inputs Protected: All Inputs Have Protection Against Static Charge**
- **Low Cost Packaging: 16 Pin Dual-In-Line Configuration**

The Intel® 2102A is a high speed 1024 word by one bit static random access memory element using N-channel MOS devices integrated on a monolithic array. It uses fully DC stable (static) circuitry and therefore requires no clocks or refreshing to operate. The data is read out nondestructively and has the same polarity as the input data.

The 2102A is designed for memory applications where high performance, low cost, large bit storage, and simple interfacing are important design objectives. *A low standby power version (2102AL) is also available. It has all the same operating characteristics of the 2102A with the added feature of 35mW maximum power dissipation in standby and 174mW in operations.*

It is directly TTL compatible in all respects: inputs, output, and a single +5 volt supply. A separate chip enable ($\overline{CE}$) lead allows easy selection of an individual package when outputs are OR-tied.

The Intel® 2102A is fabricated with N-channel silicon gate technology. This technology allows the design and production of high performance easy to use MOS circuits and provides a higher functional density on a monolithic chip than either conventional MOS technology or P-channel silicon gate technology.

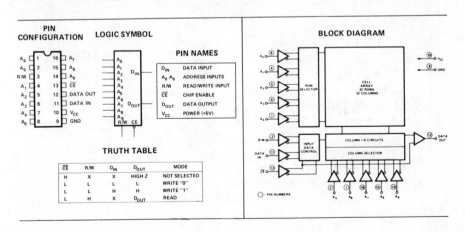

μA741
FREQUENCY-COMPENSATED OPERATIONAL AMPLIFIER
FAIRCHILD LINEAR INTEGRATED CIRCUIT

GENERAL DESCRIPTION — The μA741 is a high performance monolithic Operational Amplifier constructed using the Fairchild Planar* epitaxial process. It is intended for a wide range of analog applications. High common mode voltage range and absence of latch-up tendencies make the μA741 ideal for use as a voltage follower. The high gain and wide range of operating voltage provides superior performance in integrator, summing amplifier, and general feedback applications. Electrical characteristics of the μA741A and E are identical to MIL-M-38510/10101.

- NO FREQUENCY COMPENSATION REQUIRED
- SHORT CIRCUIT PROTECTION
- OFFSET VOLTAGE NULL CAPABILITY
- LARGE COMMON MODE AND DIFFERENTIAL VOLTAGE RANGES
- LOW POWER CONSUMPTION
- NO LATCH-UP

ABSOLUTE MAXIMUM RATINGS

Supply Voltage	
μA741A, μA741, μA741E	±22 V
μA741C	±18 V
Internal Power Dissipation (Note 1)	
Metal Can	500 mW
Molded and Hermetic DIP	670 mW
Mini DIP	310 mW
Flatpak	570 mW
Differential Input Voltage	±30 V
Input Voltage (Note 2)	±15 V
Storage Temperature Range	
Metal Can, Hermetic DIP, and Flatpak	−65°C to +150°C
Mini DIP, Molded DIP	−55°C to +125°C
Operating Temperature Range	
Military (μA741A, μA741)	−55°C to +125°C
Commercial (μA741E, μA741C)	0°C to +70°C
Lead Temperature (Soldering)	
Metal Can, Hermetic DIPs, and Flatpak (60 s)	300°C
Molded DIPs (10 s)	260°C
Output Short Circuit Duration (Note 3)	Indefinite

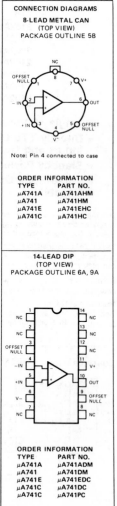

CONNECTION DIAGRAMS

8-LEAD METAL CAN
(TOP VIEW)
PACKAGE OUTLINE 5B

Note: Pin 4 connected to case

ORDER INFORMATION

TYPE	PART NO.
μA741A	μA741AHM
μA741	μA741HM
μA741E	μA741EHC
μA741C	μA741HC

14-LEAD DIP
(TOP VIEW)
PACKAGE OUTLINE 6A, 9A

ORDER INFORMATION

TYPE	PART NO.
μA741A	μA741ADM
μA741	μA741DM
μA741E	μA741EDC
μA741C	μA741DC
μA741C	μA741PC

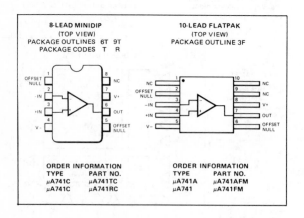

8-LEAD MINIDIP
(TOP VIEW)
PACKAGE OUTLINES 6T 9T
PACKAGE CODES   T   R

10-LEAD FLATPAK
(TOP VIEW)
PACKAGE OUTLINE 3F

ORDER INFORMATION

TYPE	PART NO.
μA741C	μA741TC
μA741C	μA741RC

ORDER INFORMATION

TYPE	PART NO.
μA741A	μA741AFM
μA741	μA741FM

FAIRCHILD LINEAR INTEGRATED CIRCUITS • μA741

μA741C

ELECTRICAL CHARACTERISTICS ($V_S = \pm 15$ V, $T_A = 25°$ C unless otherwise specified)

PARAMETERS (see definitions)		CONDITIONS	MIN	TYP	MAX	UNITS
Input Offset Voltage		$R_S \leqslant 10$ kΩ		2.0	6.0	mV
Input Offset Current				20	200	nA
Input Bias Current				80	500	nA
Input Resistance			0.3	2.0		MΩ
Input Capacitance				1.4		pF
Offset Voltage Adjustment Range				±15		mV
Input Voltage Range			±12	±13		V
Common Mode Rejection Ratio		$R_S \leqslant 10$ kΩ	70	90		dB
Supply Voltage Rejection Ratio		$R_S \leqslant 10$ kΩ		30	150	μV/V
Large Signal Voltage Gain		$R_L \geqslant 2$ kΩ, $V_{OUT} = \pm 10$ V	20,000	200,000		
Output Voltage Swing		$R_L \geqslant 10$ kΩ	±12	±14		V
		$R_L \geqslant 2$ kΩ	±10	±13		V
Output Resistance				75		Ω
Output Short Circuit Current				25		mA
Supply Current				1.7	2.8	mA
Power Consumption				50	85	mW
Transient Response (Unity Gain)	Rise time	$V_{IN} = 20$ mV, $R_L = 2$ kΩ, $C_L \leqslant 100$ pF		0.3		μs
	Overshoot			5.0		%
Slew Rate		$R_L \geqslant 2$ kΩ		0.5		V/μs

The following specifications apply for $0°$ C $\leqslant T_A \leqslant +70°$ C:

Input Offset Voltage					7.5	mV
Input Offset Current					300	nA
Input Bias Current					800	nA
Large Signal Voltage Gain		$R_L \geqslant 2$ kΩ, $V_{OUT} = \pm 10$ V	15,000			
Output Voltage Swing		$R_L \geqslant 2$ kΩ	±10	±13		V

EQUIVALENT CIRCUIT

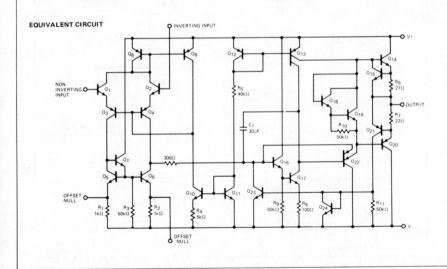

Operational Amplifiers

LM301A operational amplifier
general description

The LM301A is a general-purpose operational amplifier which features improved performance over the 709C and other popular amplifiers. Advanced processing techniques make possible an order of magnitude reduction in input currents, and a redesign of the biasing circuitry reduces the temperature drift of input current.

This amplifier offers many features which make its application nearly foolproof: overload protection on the input and output, no latch-up when the common mode range is exceeded, freedom from oscillations and compensation with a single 30 pF capacitor. It has advantages over internally compensated amplifiers in that the compensation can be tailored to the particular application. For

example, as a summing amplifier, slew rates of 10 V/μs and bandwidths of 10 MHz can be realized. In addition, the circuit can be used as a comparator with differential inputs up to ±30V; and the output can be clamped at any desired level to make it compatible with logic circuits.

The LM301A provides better accuracy and lower noise than its predecessors in high impedance circuitry. The low input currents also make it particularly well suited for long interval integrators or timers, sample and hold circuits and low frequency waveform generators. Further, replacing circuits where matched transistor pairs buffer the inputs of conventional IC op amps, it can give lower offset voltage and drift at reduced cost.

schematic** and connection diagrams

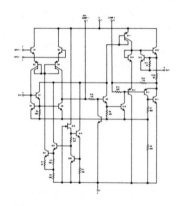

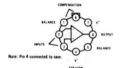

Note: Pin 4 connected to case.

Order Number LM301AH
See Package 11

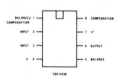

Order Number LM301AN
See Package 20

typical applications **

Integrator with Bias Current Compensation

Low Frequency Square Wave Generator

Voltage Comparator for Driving DTL or TTL Integrated Circuits

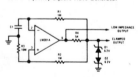

*Adjust for zero integrator drift. Current drift typically 0.1 nA/°C over 0°C to 70°C temperature range.

**Pin connections shown are for metal can.

absolute maximum ratings

Supply Voltage	±18V
Power Dissipation (Note 1)	500 mW
Differential Input Voltage	±30V
Input Voltage (Note 2)	±15V
Output Short-Circuit Duration (Note 3)	Indefinite
Operating Temperature Range	$0°C$ to $70°C$
Storage Temperature Range	$-65°C$ to $150°C$
Lead Temperature (Soldering, 10 sec)	$300°C$

electrical characteristics (Note 4)

PARAMETER	CONDITIONS	MIN	TYP	MAX	UNITS
Input Offset Voltage	$T_A = 25°C$, $R_S \leq 50\ k\Omega$		2.0	7.5	mV
Input Offset Current	$T_A = 25°C$		3	50	nA
Input Bias Current	$T_A = 25°C$		70	250	nA
Input Resistance	$T_A = 25°C$	0.5	2		MΩ
Supply Current	$T_A = 25°C$, $V_S = ±15V$		1.8	3.0	mA
Large Signal Voltage Gain	$T_A = 25°C$, $V_S = ±15V$ $V_{OUT} = ±10V$, $R_L \geq 2\ k\Omega$	25	160		V/mV
Input Offset Voltage	$R_S \leq 50\ k\Omega$			10	mV
Average Temperature Coefficient of Input Offset Voltage			6.0	30	$\mu V/°C$
Input Offset Current				70	nA
Average Temperature Coefficient of Input Offset Current	$25°C \leq T_A \leq 70°C$ $0°C \leq T_A \leq 25°C$		0.01 0.02	0.3 0.6	nA/$°C$ nA/$°C$
Input Bias Current				300	nA
Large Signal Voltage Gain	$V_S = ±15V$, $V_{OUT} = ±10V$ $R_L \geq 2\ k\Omega$	15			V/mV
Output Voltage Swing	$V_S = ±15V$, $R_L = 10\ k\Omega$ $R_L = 2\ k\Omega$	±12 ±10	±14 ±13		V V
Input Voltage Range	$V_S = ±15V$	±12			V
Common Mode Rejection Ratio	$R_S \leq 50\ k\Omega$	70	90		dB
Supply Voltage Rejection Ratio	$R_S \leq 50\ k\Omega$	70	96		dB

Note 1: For operating at elevated temperatures, the device must be derated based on a $100°C$ maximum junction temperature and a thermal resistance of $150°C$/W junction to ambient or $45°C$/W junction to case.

Note 2: For supply voltages less than ±15V, the absolute maximum input voltage is equal to the supply voltage.

Note 3: Continuous short circuit is allowed for case temperatures to $70°C$ and ambient temperatures to $55°C$.

Note 4: These specifications apply for $0°C \leq T_A < 70°C$, ±5V, $\leq V_S \leq ±15V$ and C1 = 30 pF unless otherwise specified.

◣◣ Voltage Comparators/Buffers

LM311 voltage comparator

general description

The LM311 is a voltage comparator that has input currents more than a hundred times lower than devices like the LM306 or LM710C. It is also designed to operate over a wider range of supply voltages: from standard ±15V op amp supplies down to the single 5V supply used for IC logic. Its output is compatible with RTL, DTL and TTL as well as MOS circuits. Further, it can drive lamps or relays, switching voltages up to 40V at currents as high as 50 mA.

features

■ Operates from single 5V supply

■ Maximum input current: 250 nA

■ Maximum offset current: 50 nA

■ Differential input voltage range: ±30V

■ Power consumption: 135 mW at ±15V

Both the input and the output of the LM311 can be isolated from system ground, and the output can drive loads referred to ground, the positive supply or the negative supply. Offset balancing and strobe capability are provided and outputs can be wire OR'ed. Although slower than the LM306 and LM710C (200 ns response time vs 40 ns) the device is also much less prone to spurious oscillations. The LM311 has the same pin configuration as the LM306 and LM710C.

schematic diagram and auxiliary circuits

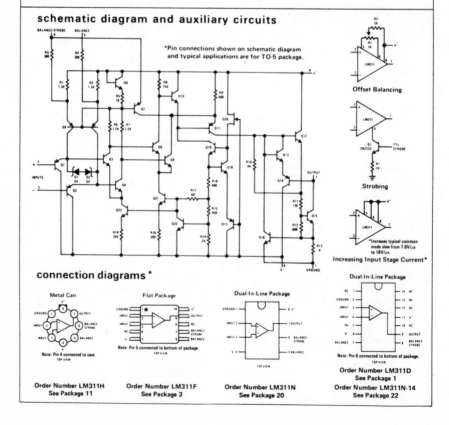

*Pin connections shown on schematic diagram and typical applications are for TO-5 package.

Offset Balancing

Strobing

Increasing Input Stage Current*

connection diagrams *

Metal Can

Note: Pin 4 connected to case.
TOP VIEW

Order Number LM311H
See Package 11

Flat Package

Note: Pin 5 connected to bottom of package.
TOP VIEW

Order Number LM311F
See Package 3

Dual-In-Line Package

TOP VIEW

Order Number LM311N
See Package 20

Dual-In-Line Package

Note: Pin 6 connected to bottom of package.
TOP VIEW

Order Number LM311D
See Package 1

Order Number LM311N-14
See Package 22

absolute maximum ratings

Total Supply Voltage (V_{84})	36V
Output to Negative Supply Voltage (V_{74})	40V
Ground to Negative Supply Voltage (V_{14})	30V
Differential Input Voltage	±30V
Input Voltage (Note 1)	±15V
Power Dissipation (Note 2)	500 mW
Output Short Circuit Duration	10 sec
Operating Temperature Range	$0°C$ to $70°C$
Storage Temperature Range	$-65°C$ to $150°C$
Lead Temperature (soldering, 10 sec)	$300°C$

electrical characteristics (Note 3)

PARAMETER	CONDITIONS	MIN	TYP	MAX	UNITS
Input Offset Voltage (Note 4)	$T_A = 25°C$, $R_S \leq 50K$		2.0	7.5	mV
Input Offset Current (Note 4)	$T_A = 25°C$		6.0	50	nA
Input Bias Current	$T_A = 25°C$		100	250	nA
Voltage Gain	$T_A = 25°C$		200		V/mV
Response Time (Note 5)	$T_A = 25°C$		200		ns
Saturation Voltage	$V_{IN} \leq -10\,mV$, $I_{OUT} = 50\,mA$ $T_A = 25°C$		0.75	1.5	V
Strobe On Current	$T_A = 25°C$		3.0		mA
Output Leakage Current	$V_{IN} \geq 10\,mV$, $V_{OUT} = 35V$ $T_A = 25°C$		0.2	50	nA
Input Offset Voltage (Note 4)	$R_S \leq 50K$			10	mV
Input Offset Current (Note 4)				70	nA
Input Bias Current				300	nA
Input Voltage Range			±14		V
Saturation Voltage	$V^+ \geq 4.5V$, $V^- = 0$ $V_{IN} \leq -10\,mV$, $I_{SINK} \leq 8\,mA$		0.23	0.4	V
Positive Supply Current	$T_A = 25°C$		5.1	7.5	mA
Negative Supply Current	$T_A = 25°C$		4.1	5.0	mA

Note 1: This rating applies for ±15V supplies. The positive input voltage limit is 30V above the negative supply. The negative input voltage limit is equal to the negative supply voltage or 30V below the positive supply, whichever is less.

Note 2: The maximum junction temperature of the LM311 is $85°C$. For operating at elevated temperatures, devices in the TO-5 package must be derated based on a thermal resistance of $150°C/W$, junction to ambient, or $45°C/W$, junction to case. For the flat package, the derating is based on a thermal resistance of $185°C/W$ when mounted on a 1/16-inch-thick epoxy glass board with ten, 0.03-inch-wide, 2-ounce copper conductors. The thermal resistance of the dual-in-line package is $100°C/W$, junction to ambient.

Note 3: These specifications apply for $V_S = \pm15V$ and $0°C < T_A < 70°C$, unless otherwise specified. The offset voltage, offset current and bias current specifications apply for any supply voltage from a single 5V supply up to ±15V supplies.

Note 4: The offset voltages and offset currents given are the maximum values required to drive the output within a volt of either supply with 1 mA load. Thus, these parameters define an error band and take into account the worst case effects of voltage gain and input impedance.

Note 5: The response time specified (see definitions) is for a 100 mV input step with 5 mV overdrive.

Operational Amplifiers

LM3900 quad amplifier

general description

The LM3900 consists of four independent, dual input, internally compensated amplifiers which were designed specifically to operate off of a single power supply voltage and to provide a large output voltage swing. These amplifiers make use of a current mirror to achieve the non-inverting input function. Application areas include: AC amplifiers, RC active filters; low frequency triangle, squarewave and pulse waveform generation circuits, tachometers and low speed, high voltage digital logic gates.

features
- Wide single supply voltage range   4 V_{DC} to 36 V_{DC}
  or dual supplies   ±2 V_{DC} to ±18 V_{DC}
- Supply current drain independent of supply voltage
- Low input biasing current   30 nA
- High open-loop gain   70 dB
- Wide bandwidth   2.5 MHz (Unity Gain)
- Large output voltage swing $(V^+ - 1)$ V_{p-p}
- Internally frequency compensated for unity gain
- Output short-circuit protection

schematic and connection diagrams

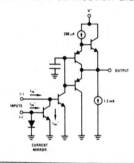

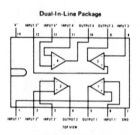

Dual-In-Line Package

Order Number LM3900N
See Package 22

PARAMETER	CONDITIONS	MIN	TYP	MAX	UNITS
Open Loop					
Voltage Gain	f = 100 Hz	1200	2800		V/V
Input Resistance	Inverting Input		1		MΩ
Output Resistance			8		kΩ
Unity Gain Bandwidth	Inverting Input		2.5		MHz
Input Bias Current	Inverting Input		30	200	nA
Slew Rate	Positive Output Swing		0.5		V/µs
	Negative Output Swing		20		V/µs
Supply Current	$R_L = \infty$ On All Amplifiers		6.2	10	mA DC
Output Voltage Swing	$R_L = 5.1k$				
V_{OUT} High	$I_{IN^-} = 0, I_{IN}^+ = 0$	13.5	14.2		VDC
V_{OUT} Low	$I_{IN^-} = 10\,\mu A, I_{IN}^+ = 0$		0.09	0.2	VDC
Output Current Capability					
Source		3	18		mA DC
Sink	(Note 2)	0.5	1.3		mA DC
Power Supply Rejection	f = 100 Hz		70		dB
Mirror Gain	$I_{IN}^+ = 200\,\mu A$ (Note 3)	0.9	1	1.1	µA/µA
Mirror Current	(Note 4)		10	500	µA DC
Negative Input Current	(Note 5)		1.0		mA DC

Voltage Regulators

LM340 series 3-terminal positive regulators

general description

The LM340-XX series of three terminal regulators is available with several fixed output voltages making them useful in a wide range of applications. One of these is local on card regulation, eliminating the distribution problems associated with single point regulation. The voltages available allow these regulators to be used in logic systems, instrumentation, HiFi, and other solid state electronic equipment. Although designed primarily as fixed voltage regulators these devices can be used with external components to obtain adjustable voltages and currents.

The LM340-XX series is available in two power packages. Both the plastic TO-220 and metal TO-3 packages allow these regulators to deliver over 1.0A if adequate heat sinking is provided. Current limiting is included to limit the peak output current to a safe value. Safe area protection for the output transistor is provided to limit internal power dissipation. If internal power dissipation becomes too high for the heat sinking provided, the thermal shutdown circuit takes over preventing the IC from overheating.

Considerable effort was expended to make the LM340-XX series of regulators easy to use and minimize the number of external components. It is not necessary to bypass the output, although this does improve transient response. Input bypassing is needed only if the regulator is located far from the filter capacitor of the power supply.

features

- Output current in excess of 1A
- Internal thermal overload protection
- No external components required
- Output transistor safe area protection
- Internal short circuit current limit
- Available in plastic TO-220 and metal TO-3 packages

voltage range

LM340-05	5V	LM340-15	15V
LM340-06	6V	LM340-18	18V
LM340-08	8V	LM340-24	24V
LM340-12	12V		

schematic and connection diagrams

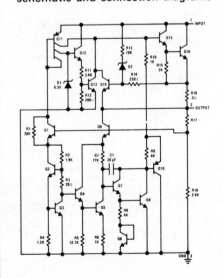

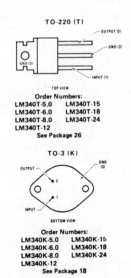

TO-220 (T)

TOP VIEW

Order Numbers:

LM340T-5.0   LM340T-15
LM340T-6.0   LM340T-18
LM340T-8.0   LM340T-24
LM340T-12
See Package 26

TO-3 (K)

BOTTOM VIEW

Order Numbers:

LM340K-5.0   LM340K-15
LM340K-6.0   LM340K-18
LM340K-8.0   LM340K-24
LM340K-12
See Package 18

absolute maximum ratings

Input Voltage (V_O = 5V through 18V)	35V
(V_O = 24V)	40V
Internal Power Dissipation (Note 1)	Internally Limited
Operating Temperature Range	$0°C$ to $70°C$
Maximum Junction Temperature	
TO-3 Package	$150°C$
TO-220 Package	$150°C$
Storage Temperature Range	$-65°C$ to $+150°C$
Lead Temperature	
TO-3 Package (Soldering, 10 sec)	$300°C$
TO-220 Package (Soldering, 10 sec)	$230°C$

electrical characteristics

LM340-5 (V_{IN} = 10V, I_{OUT} = 500 mA, $0°C \leq T_A \leq 70°C$, unless otherwise specified)

PARAMETER	CONDITIONS	MIN	TYP	MAX	UNITS
Output Voltage	$T_j = 25°C$	4.8	5	5.2	V
Line Regulation	$T_j = 25°C$, $7V \leq V_{IN} \leq 25V$				
	I_{OUT} = 100 mA			50	mV
	I_{OUT} = 500 mA			100	mV
Load Regulation	$T_j = 25°C$, $5 mA \leq I_{OUT} \leq 1.5A$			100	mV
Output Voltage	$7V \leq V_{IN} \leq 20V$, $5 mA \leq I_{OUT} \leq 1.0A$	4.75		5.25	V
	$P_D \leq 15W$				
Quiescent Current	$T_j = 25°C$		7	10	mA
Quiescent Current Change	$7V \leq V_{IN} \leq 25V$			1.3	mA
	$5 mA \leq I_{OUT} \leq 1.5A$			.5	mA
Output Noise Voltage	$T_A = 25°C$, $10 Hz \leq f \leq 100 kHz$		40		µV
Long Term Stability				20	mV/1000 hr
Ripple Rejection	f = 120 Hz		60		dB
Dropout Voltage	$T_j = 25°C$, I_{OUT} = 1.0A		2		V

LM340-15 (V_{IN} = 23V, I_{OUT} = 500 mA, $0°C \leq T_A \leq 70°C$, unless otherwise specified)

PARAMETER	CONDITIONS	MIN	TYP	MAX	UNITS
Output Voltage	$T_j = 25°C$	14.4	15	15.6	V
Line Regulation	$T_j = 25°C$, $17.5V \leq V_{IN} \leq 30V$				
	I_{OUT} = 100 mA			150	mV
	I_{OUT} = 500 mA			300	mV
Load Regulation	$T_j = 25°C$, $5 mA \leq I_{OUT} \leq 1.5A$			300	mV
Output Voltage	$17.5V \leq V_{IN} \leq 30V$, $5 mA \leq I_{OUT} \leq 1.0A$	14.25		15.75	V
	$P_D \leq 15W$				
Quiescent Current	$T_j = 25°C$		7	10	mA
Quiescent Current Change	$17.5V \leq V_{IN} \leq 30V$			1	mA
	$5 mA \leq I_{OUT} \leq 1.5A$			.5	mA
Output Noise Voltage	$T_A = 25°C$, $10 Hz \leq f \leq 100 kHz$		90		µV
Long Term Stability				60	mV/1000 hr
Ripple Rejection	f = 120 Hz		50		dB
Dropout Voltage	$T_j = 25°C$, I_{OUT} = 1.0A		2		V

Appendix B

An alphabetical list of the functions performed by integrated circuits

DIGITAL INTEGRATED CIRCUITS

A

Accumulators
Adders
  binary
  full
  half
ALU (arithmetic logic unit)
Arithmetic operators

B

Buffers

C

Calculators
CLA (carry look ahead)
Clocks
  digital
  digital alarm
  driver
  generator

Comparators
  bus
  differential
  high speed
  magnitude
  n bit
  voltage
Controllers
Convertors
  BCD to binary
  binary to BCD
  code
  level
  parallel to serial
  serial to parallel
Counter/latch
Counters
  asynchronous
  binary
  decade
  divide by n

Switches
  debouncers
  digital
Synthesizers

T

Timekeepers
Transceivers
Translators, voltage

LINEAR INTEGRATED CIRCUITS

A

Amplifiers
  AGC/squelch
  audio
  buffer
  chroma
  current
  current-mode
  differential
  instrumentation
  logarithmic
  low noise
  low power
  operational
  power
  preamplifiers
  programmable
  RF/IF
  sense
  video
  wideband
Analog switches
AM
  IF amplifier/detector
  IF strip
  radio system
  RF/IF amplifiers
  video amplifier/detector

C

Comparators
  differential
  high speed
  low power
  voltage

Controllers
Converters
  analog to digital
  digital to analog
  frequency to voltage
  voltage to frequency

D

Demodulators
Detectors
  fluid
  level
Diode arrays
Drivers
  clock
  line
  memory
  peripheral

F

FM
  audio amplifier
  detector/limiter
  IF system
  multiplex stereo demodulator

L

Level detectors
Limiters

M

Mixers
Modems
Modulator/demodulator

O

Oscillators, voltage controlled

P

Phase locked loops

R

Receivers, line
References, voltage
Regulators
    fixed
    negative voltage
    positive voltage
    series
    variable

S

Sample and hold
Sources
    current
    voltage

T

Timers
Tone decoders
Transceivers
Transducers
    pressure
    temperature
Transistor arrays
TV
    AFT (automatic fine tuning)
    chroma demodulator
    chroma IF strip
    chroma processor
    chroma subcarrier regenerator
    sound system
    video, IF, and detector
    video modulator

V

VCO (voltage controlled oscillator)

Index